5/07

BUILDING
A SUCCESSFUL
CONSTRUCTION
COMPANY

PATRICIA W. ATALLAH

KAPLAN AEC EDUCATION

This publication is designed to provide accurate and authoritative information in regard to the subject matter covered. It is sold with the understanding that the publisher is not engaged in rendering legal, accounting, or other professional service. If legal advice or other expert assistance is required, the services of a competent professional should be sought.

President, Kaplan Publishing: Roy Lipner
Vice President and Publisher: Maureen McMahon
Acquisitions Editor: Victoria Smith
Development Editor: Trey Thoelcke
Production Editor: Leah Strauss
Typsetter: the dotted i
Cover Designer: Sue Giroux

Published by Kaplan Publishing,
a division of Kaplan, Inc.

Printed in the United States of America

06 07 08 10 9 8 7 6 5 4 3 2 1

Library of Congress Cataloging-in-Publication Data

Atallah, Patricia W.
  Building a successful construction company : create a strategy, organize your business, protect your bottom line / Patricia W. Atallah.
    p. cm.
  Includes bibliographical references and index.
  ISBN-13: 978-1-4195-2811-8
  ISBN-10:   1-4195-2811-4
  1. Construction industry—Management. 2. New business enterprises—Management. I. Title.
  HD9715.A2A85 2006
  624.068'4–dc22

                                                                    2006012336

Kaplan Publishing books are available at special quantity discounts to use for sales promotions, employee premiums, or educational purposes. Please call our Special Sales Department to order or for more information at 800-621-9621, ext. 4444, e-mail kaplanpubsales@kaplan.com, or write to Kaplan Publishing, 30 South Wacker Drive, Suite 2500, Chicago, IL 60606-7481.

Contents

PROTECT YOUR BOTTOM LINE

This is the first book I've written and it could not have been started and completed without the guidance, contributions, and support of many people. I'd like to thank everyone for making this a truly wonderful and rewarding experience for me.

At the top of my list is Victoria Smith, my editor at Kaplan Publishing, and the Kaplan team. Call it fate or just good timing, Victoria received my book proposal just as she was about to launch a new book division dedicated to the architectural, engineering, and construction industries. From that moment on, she has provided steady guidance and support (not to mention understanding, patience, and flexibility!). Lisa Schuble, associate publicist, and April Timm, corporate marketing manager, round out a highly professional and responsive team at Kaplan.

Next, I'd like to thank the industry experts and company CEOs who took precious time out of their schedules to meet with me and share their knowledge, insights, "war stories" and passion for the industry. These meetings were the real high points of my book-writing experience. I'd like to give special thanks to Jay Several (ETRAC Solutions, Inc.) for his special contribution to Chapter 6. And, in alphabetical order, thanks to: Scott Adams (Avalon Risk Associates); Jim Anchin (Anchin Block & Anchin LLP); Craig Belesi; Dennis Chamberlain (St. Paul Travelers); Diane Cramer and Barry Fries (B.R. Fries); Henry Goldberg (Goldberg & Connolly); Alan Green (Wachovia Bank); Bill Haas (USI Holdings Corp.); Hank Harris (FMI Corp.); Keith Housman (Housman & Bloch, LLP); Eric Kreuter, PhD (Marden, Harrison & Kreuter CPAs, P.C.); Peter Lehrer (cofounder, Lehrer McGovern); Fred Levinson (Levinson & Santoro Electric); Bruce Lindenbaum (Frank & Lindy Plumbing); Arnold Marden (Marden, Harrison & Kreuter CPAs, P.C.); Jim McKenna (Hunter Roberts Construction); Michele Medaglia (ACC Construction); Kirk Ortega (Ortega Group); Jack Osborn (John E. Osborn, P.C.); Scott Rives (Woodworks Construction Co.); Tom Rogers

(Signature Bank); Danny Sawh (Stonewall Contracting); Frank Sciame (F.J. Sciame); Lou Silbert (consultant); Ken Simonson (Associated General Contractors of America); Marcelo Velez (Columbia University); Terry Yeager (Navigant Consulting); Michael Zetlin (Zetlin & De Chiara LLP); Jeffrey Zogg (General Building Contractors of New York, AGC Chapter).

Last but not least, I'd like to thank my husband, Akram—partner in life and work—for his constructive ideas and feedback, unwavering support and patience; my children, Alexander and Juliana, for their quiet tolerance as I sat hunched over my laptop for hours on end; my friends and my sister, Bianca Wisznat, for their cheerleading.

If you're seriously thinking about starting a contracting business in the construction industry, or are already deep into it and wake up in a cold sweat nightly wondering how you'll get through the week, this book is for you. I've written it with you in mind because I've "been there" and know firsthand how it feels to grope one's way through the process of starting and growing a business and then stay afloat and even thrive in a tough and quirky industry like construction. I've also worked with an assortment of contractors over the years and have gained insights on what factors separate success from failure. I'd like to share those with you here.

This book is designed specifically for aspiring and start-up contractors who want to get it right from the start, and for established contractors who are frustrated and overwhelmed with their businesses and tired of operating in a crisis management mode. If either of these describes you, this book will help you take a breather from your daily routine and walk you through the steps you must take to start thinking "big picture" about your business. Only then can you properly assess your current situation and the challenges and opportunities ahead of you. Only then can you decide where you want to go and how you are going to get there.

I started a construction business more than 12 years ago with business and banking experience and scant knowledge of the construction industry. What on earth possessed me, you ask? I've always had an entrepreneurial bent, and in my early 30s, I became anxious to drop out of the corporate fold and start my own business. I was looking for flexibility, a better balance in my life, and freedom from the limitations of a job description. I researched various possibilities for about a year and, based on my research, finally decided to start a business in the construction industry. With the perspective of an outsider looking in, I recognized some of the critical issues facing the industry and saw an opportunity to eventually make a contribution.

I joined forces with two partners who were trained engineers working for a prominent construction company. We looked at industry trends and the outlook for the mid-1990s and beyond, and envisioned that we could achieve great things with our combined talents. I was excited by the prospect of launching a business in a new industry and was attracted and comforted by the notion that I wouldn't be venturing out alone. Our target markets were commercial and institutional general contracting and construction consulting.

And so we plunged in with boundless enthusiasm and incredible focus. With virtually no capital and few resources, we managed to "move mountains" by winning projects and building name recognition in a mature and crowded market like New York. Beginner's luck? High energy fueled by unbeatable optimism? Yes to both.

In the middle to late 1990s, we experienced several ups and downs, at least one major crisis, several close calls, and almost daily thrills and disappointments. Despite our impressive credentials and our belief that we knew what we were doing, we made painful mistakes early on and almost went under in 1997, during our third year.

Looking back, we refer to those hard lessons as the "most expensive degree we ever earned" and a turning point in our business. We decided to get smart and establish realistic and well-defined goals and plans for our business instead of "playing it by ear." Since the late 1990s, we've narrowed our focus to construction consulting and have steadily built a professional organization with carefully targeted services, a team of top professionals, a solid client base, and a market presence that should serve us well in the foreseeable future. The past fall, we celebrated our 12th year and are looking forward to many more.

One area of consulting that I've devoted my attention to is to provide management advisory services to construction companies. Armed with a real appreciation for the challenges building contractors face every day, I've developed practical tools and advisory services aimed at helping contractors avoid the difficulties that we encountered. During the past eight years, I've helped many small and midsize construction firms take time out to look inward, assess their goals and capabilities, position themselves properly in the market, and resolve internal issues that get in the way of success.

What I've learned from working with these companies is that the construction industry is relatively easy to get into but a tough environment in which to do business, as evidenced by one of the highest business failure rates among major industries. Construction is a complex and risky activity with unknown variables and hazards at every turn. It's also an industry with constantly shifting market conditions and demands. The risks are enormous and business stability and financial rewards remain elusive for most contractors.

Faced with a variety of challenges that can affect their businesses, owners of construction companies often have trouble juggling all the issues that arise because they must manage the business and manage the project work simultaneously, a tall order for most people.

The majority of contractors know their trade and/or are proficient technically but are often weak on business management and leadership skills. Thus, they don't have the proper internal company setup and framework to allow them to manage their business risks successfully. They struggle to survive year in and year out and remain frustrated because they're not earning an acceptable return for their efforts.

If you're an owner of a start-up or a growing construction company, you're likely enthusiastic about your business and ready to do whatever it takes to succeed. You know from experience and research what some of the pitfalls are but refuse to imagine that you could end up becoming a member of the struggling majority. You're determined to do it right.

If you're an owner of an established company that is perpetually living in crisis management and survival modes, you may be fed up enough to want to make the necessary changes to put your company back on the right track.

I wrote this book to help you move ahead with your plans to succeed. To take the driver's seat, you'll need to face the realities of the business you're in and create the strategy, organizational structure, controls, and support required to propel your company toward long-term success and financial stability.

In the next ten chapters, we're going to concentrate on the key ingredients to business success: big picture thinking, strong leadership, and prudent business planning and management. I provide hard facts, true stories, practical advice, and valuable resources to help you develop and implement the strategic business thinking and management skills that are absolutely essential to your business.

We'll take a close look at the realities, risks, and rewards of the construction industry and whether they truly suit your capabilities, resources, and disposition. Next, we'll delve into the important elements of a successful business: careful planning, targeted marketing, effective company organization, strong in-house talent and resources, standard company practices and procedures, effective project management, proper financial management, prudent protection against risk, and adaptability to a changing environment.

By reading about these topics and taking advantage of the tools and resources I've made available throughout the book, you'll learn how to gain the competitive edge that is so critical to your success, insulate your company from the ebbs and flows of the market, and manage your risk. You'll also develop a real appreciation for what it takes and start laying out a workable plan to achieve your goals.

This book is designed to be user-friendly, with links to essential industry resources and checklists, templates, and workbooks that have been specifically created for your convenience. You can retrieve digital copies of many of the materials in the book by visiting the book Web site, *www.ConstructBiz.com.*

While writing this book, I had the pleasure of interviewing many contractors at various stages of their companies' development. I also met with project owners and top industry experts in the fields of management consulting, banking, accounting, financial management, bonding, law, and information systems. Throughout the book, I'll be sharing their comments, feedback, anecdotes, and wisdom.

The chapters are organized like building blocks. Each chapter builds on the preceding chapters and is the foundation for the next one. To get the most out of the book, I recommend that you read the chapters in sequence, as if you were reading a novel. You'll meander through the twists and turns of the plot (creating a successful contracting business) and hopefully find a treasure of ideas, insights, strategies, and tools that will prove invaluable as you map out your path to success.

1

LOOK BEFORE YOU LEAP

If you're in the thick of starting and growing a contracting business, it's likely that you have a lot of energy, optimism, and perseverance. Your desire for independence is overwhelming, and you're eager to make your dreams come true. You want to be your own boss and control your destiny. You probably have prior experience working for a contractor and possess specific trade or project management skills. It's your craft and hopefully your passion, and you believe it can be made into a lucrative business. You may have met contractors who have done well over the years and who enjoy excellent reputations, steady repeat business, and exciting prospects for growth. You believe that all it takes is a strong market and a couple of contracts to propel you forward.

Great attitude, but you would be wise to take a reality check. Let's review the cold hard facts about the industry and the risks and challenges that make it a tough environment in which to conduct business.

WHAT ARE YOU GETTING YOURSELF INTO?

The construction industry is one of the largest in the United States, consisting of 12 percent of all businesses, and is by far the largest construction market in the world. A major source of jobs (with 7.3 million employees), it makes a large contribution to the economy. The value of construction put in place in 2005 exceeded $1.1 trillion, nearly 9 percent of the gross domestic product (GDP).[1] The industry is diversified among public- and private-sector residential, commercial, industrial, heavy construction, and specialty trades. In 2005, private residential construction accounted for 56 percent of work put in place, while private nonresidential and public construction each accounted for 22 percent.[2]

Unlike many other industries (e.g., consumer products, pharmaceuticals, automobiles), construction is highly fragmented, with mostly small firms operating in their local geographic markets and only a relative handful of large companies with regional or national reach. According to the U.S. Census Bureau, more than 2.8 million construction concerns were in business in 2003, the vast majority of which were very small: approximately 2.1 million establishments were self-employed individuals without paid employees; the remaining 732,000 construction firms were mostly small, privately held businesses with an average of nine employees. Only 8 percent of all construction firms had more than 20 employees and a mere 1 percent had 100 or more employees![3]

The industry also differs from most others because every construction project presents a unique situation and set of challenges. So, although the end result is always a finished product (e.g., a building, bridge), the processes of execution may differ. Construction is also people-intensive, requiring daily cooperation and communication between various players, including the customer, architect and engineers, the prime contractor and subcontractors, manufacturers and vendors, and government and regulatory agencies. These complexities lead to inefficiencies at every stage of the process and combative relationships between the parties.

Michael E. Porter, professor at the Harvard Business School and a leading authority on competitive strategy, names five key determinants to an industry's potential for profitability: the extent of barriers to entry; the ease of product substitution; the degree of competition; the level of bargaining power with the customer; and the level of bargaining power with the supplier. Based on the major characteristics of the construction industry—low barriers to entry, commodity-like services, and weak bargaining power—Porter's model shows that the industry is prone to low profitability and growth.[4]

The facts speak for themselves. The Internal Revenue Service reported that, in 2002, only 60 percent of all construction companies were profitable, with an average net income of 2.6 percent of total receipts.[5] Moreover, every year, about as many new contractors start up as go out of business. According to the U.S. Small Business Administration Office of Advocacy, the business failure rate in 2002 was more than 13 percent.[6] As one contractor described it, construction presents "the possibility for a great beginning and a terrible end."

This begs the question, "Where's the learning curve?" Paraphrasing Albert Einstein, Are we caught perpetually doing the same thing and hoping for different results? Looking at the statistics, it would appear to be the case. And yet, many construction companies of all sizes and from all segments of the industry grow and prosper in their local markets. Some even expand to become national and international industry leaders. I surmise that their success hinges on the ability to properly appreciate and manage the business risks and challenges they face.

So, before we go any further in the book, let's review the major risks and how you can tackle them head-on. In Chapters 2 through 10, I will roll out specific approaches and strategies to effectively manage them.

The Risks

Ease of Entry

Construction is relatively easy to get into. It's not mandatory to have a specific degree or training and, in many states, contractors don't need to be licensed unless they represent a specific trade (e.g., electrician, plumber). The result is that many people, from all walks of life and stages in their careers, with varying skills, are attracted by the ease of entry and the perceived allure of making a lot of money. They plunge in, oftentimes not fully aware of the business risks involved.

Action: *Complete the Self-Assessment Questionnaire near the end of this chapter to confirm that you have the disposition, capabilities, and resources required to run a construction business.*

Price Competition

Most contractors don't offer a unique product or service that would give them a leg up over competition. Consequently, they compete against a multitude of similar firms for the work and win more often based on price than on quality or ability to deliver a project on time. In mature markets, too many contractors chase the same projects. Hank Harris, president of FMI Corporation, the nation's largest management consulting and investment banking firm specializing in the construction industry, refers to the competitive situation as "a car crash in the parking lot of the customer." In Darwinian terms, "the fittest will survive."

Action: *Position your company in the market so you're not competing solely on the basis of price.*

Low Profitability

In a slow market, contractors tend to underbid to get the work and accept low to nonexistent returns that don't adequately cover their costs and risks. In a strong market, prices escalate because of shortages in supervisory personnel, labor, and materials. Contractors may start out with a healthy target profit, but increased costs and other circumstances, some out of the contractor's control, will often lead to "profit fade" or a downright loss. Unlike the high risk/high return, low risk/low return scenarios we expect in financial markets, little to no correlation exists between risk and reward in construction.

Action: *Build an equity base that can sustain you through the ups and downs and keep you ahead of the "desperation curve."*

Supply Chain Inefficiencies

The average contractor isn't in control of labor and materials costs. He or she doesn't have any significant buying power and can't take advantage of the economies of scale in procurement, and ends up passing on the higher costs to the customer.

Action: *Establish good relationships with your subcontractors, vendors, and suppliers.*

Volatile Material Costs

In recent years, construction material costs have fluctuated (mostly risen) because of increases in fuel costs, shortages of materials (e.g., cement, concrete, steel), and increased demand in certain sectors (e.g., gypsum and wood products used in single-family and multifamily housing). The U.S. Department of Labor released data showing that the cost of building a warehouse rose 7.6 percent in 2005, well above the consumer inflation rate of 3.4 percent. Contractors bound to a fixed-price contract can find themselves liable for these increases, especially on public-sector projects.[7]

Action: *Be careful how you bid on projects and how your contract reads regarding cost escalations.*

Changing Market Conditions

Construction is a cyclical business. Even if the local market is hot now, a downturn is bound to occur as one area becomes saturated with certain types of construction, or as activity slows with the economy. Each time there is a trough in the market, an alarming number of contractors, small and large, go out of business.

Action: *Stay up-to-date on industry and economic trends and outlook.*

Geographic Constraints

Construction is mostly a provincial business in which contractors respond to regional market conditions and rely on local subcontractors and suppliers to participate in the work. Each market has its own quirks and requirements; when contractors decide to expand beyond their "borders," they're exposing themselves to new rules and unknown risks.

Action: *Take the time to make prudent bid decisions; avoid bidding on work in unfamiliar areas.*

Shortages of Qualified Management and Labor

Good people are hard to find when you need them most, especially in the construction industry. Try hiring qualified project managers while the market is on the upswing! Even in a slow market, however, management talent is a precious commodity because, according to Hank Harris of FMI, the industry has not placed enough emphasis on management recruiting, training, and development. Finding and keeping quality skilled labor is also a challenge, especially when times are good and the labor market is tight. I recall the shortage of labor experienced in New York City during the "boom years" of the late 1990s, and how carpenters, electricians, and other skilled tradespeople arrived in droves from other states.

Action: *A construction company is only as good as the people who work there. Make it a top priority to attract and retain good people.*

Complexity of the Work

Projects are becoming increasingly complex because of various factors, including fluctuating economic conditions, limited funding, changes in environmental and safety requirements, and other regulations. Property owners and developers have become more demanding, requiring contractors to work under tight budgets, accelerated schedules, and difficult site conditions, and to demonstrate high competency in specialty fields. Contractors must possess a level of sophistication in skills and resources that goes well beyond expectations of even five years ago.

Action: *Bid on projects you are qualified for; build a team of competent people; make sure you keep pace with new methodologies and technologies.*

Wide Spectrum of Project Delivery Alternatives

Over the years, the construction industry has sprouted a variety of "project delivery systems," the definitions of which have become ever more nebulous. Project owners face a confusing spectrum of choices: do they hire a "general contractor," "construction manager," "construction manager at risk," "builder," "design/builder," or some derivative or combination of these? To compete successfully, the contractor must fully understand the various permutations evolving in the market and how his or her company should be defined.

Action: *Understand the needs of the market; define your business and send a clear message about your services.*

Tight Financing

Deep pockets and willing backers are rare in the construction industry. A contractor is unlikely to obtain a bank credit line unless he or she has a track record and sufficient personal assets. Consequently, most contractors start off with virtually no capital or access to credit and bonding. They keep their fingers crossed, hoping to win big on the first project. With limited resources, they stumble the first time they fall short of their cash-flow needs and the daily headaches begin. Running out of money to make payroll and pay suppliers in the middle of a project can be the death of a contractor.

Action: *Make sure you have cash available to start your business and build it slowly, brick by brick.*

Tough Bonding and Insurance Environments

Small to medium-size contractors are finding it more difficult than ever to obtain the bonding required to perform public projects and to find affordable insurance products. Huge losses suffered by the insurance industry in recent years have caused many of them to discontinue their bonding business or to shrink their portfolios by being more selective. They've also raised their insurance premiums to unprecedented levels, making it difficult for the small contractor to afford basic insurance.

Action: *Find a good construction accountant and bonding agent who can facilitate the process of obtaining bonding and insurance.*

Dependence on Others

Contractors find themselves at the mercy of others to complete their projects successfully, whether it is the owner or prime contractor for payment, the bank for sufficient working capital funding, or subcontractors and suppliers for satisfactory work performance.

Action: *Establish a collaborative relationship of give-and-take with project participants.*

Low Trust Factor

Most contractors are honest, hardworking people who follow the rules. However, the "bad apples" who don't deliver as promised often give the whole lot a bad rap. Property owners and other project participants tend to be wary, which often leads to an adversarial atmosphere on a project, which, in turn, puts the contractor on the defensive and keeps construction lawyers busy.

Action: *Be sure to convey to your customers the values your company embraces.*

Uncontrollable Variables

Curveballs are thrown at contractors all the time—canceled or delayed contracts, mistakes made on a bid, unforeseen site conditions, accidents and injuries, bad weather, scarce labor and materials, fluctuations in prices, poor performance by subcontractors and suppliers, increases in insurance requirements and premiums, uncooperative project participants, a difficult or unreasonable customer, insufficient project funding, and on and on. These surprises can lead to a potential loss or an all-out disaster.

Action: *Avoid seat-of-the-pants crisis management; establish a framework for planning and executing your projects.*

The Challenges

Faced with a variety of challenges that can affect their businesses, construction company owners often have trouble juggling all the issues that arise on a daily basis. Why is it so difficult? The answer is two-pronged:

1. Most are proficienct in a particular trade or technical discipline but have little or no experience in leading a business operation.
2. All are confronted by two sets of management challenges that are distinct but interrelated: managing the business and managing the project work.

Like it or not, you must deal with the same strategic issues as any other business owner, including:

- Leading the organization
- Managing cash flow
- Obtaining bank financing, bonding, and insurance
- Finding and keeping good management and office personnel
- Providing excellent customer service
- Identifying and controlling costs
- Improving profitability
- Staying focused on company goals
- Establishing and implementing a marketing plan
- Managing company growth or expansion

In addition, your construction projects bring a unique bundle of management issues that, if not dealt with properly, can lead to nonperformance and painful financial or even human loss. These are:

- Bidding accurately on projects and controlling costs
- Getting the work
- Managing subcontractors' performance
- Managing procurement efficiently

- Handling labor issues
- Dealing with the responsibility for safety
- Managing contract issues
- Controlling labor efficiency and progress
- Finding good field people
- Complying with government policies and requirements
- Handling cumbersome paperwork and reporting requirements
- Pleasing the customer
- Working effectively with other project participants
- Avoiding disputes

It's a daily challenge to develop and maintain the internal organization, skills, and stamina required to manage all these activities effectively. As one contractor put it, "You've got to constantly work to stay ahead of the desperation curve." Remember, one bad project can wreak havoc on your entire operation and even lead to business failure.

The result is that you may be struggling to survive year in and year out. You may be easily overwhelmed and frustrated because you're operating in a crisis mode 24/7 and aren't earning an acceptable return for your efforts. You're likely to be so busy putting out fires that your business will suffer, which, in turn, eventually leads to other fires down the road. One contractor told me, "It's like I'm on a roller-coaster ride and can't ever get off!" Another described it similarly as a "treadmill gone haywire." Nerves become frayed, patience runs thin, and burnout takes over. That's obviously not where you want to be!

WHAT DOES IT TAKE?

As a business owner, your goal is simple: at a minimum, you want to win new and repeat business that can sustain your operation and eventually lead to long-term stability and success. To accomplish this, you must deliver top-quality and efficient service, at optimal cost, to the satisfaction of your customers.

Achieving your goal, however, is no slam dunk. It takes a great deal of planning, hard work, and trial and error to start, grow, and stabilize a business, deliver consistently, and survive in the long haul. The hurdles are everywhere and can lead to chronic headaches and sleepless nights.

Contracting 101: Planning, Management, and Controls Are KEY

Back in 1996, my partners and I decided to bid on a school project as general contractor on a lump-sum basis. With all of our sophistication and knowledge, we knew little about the practical realities of the process, and it was almost a disaster. The first mistake we made was to plunge into something without real hands-on experience. The second was that we underbid the project by forgetting to include site supervision costs! Third, to make matters worse, we began the work without a clear-cut plan of execution. To please the client, we submitted a project schedule that reflected exactly what the school wanted even though we believed it was too ambitious. We also didn't have strong ties to subcontractors and discovered that it was difficult to control them. Then the rains came and literally washed the project out for more than two months, which caused a chain reaction of problems and a bit of antagonism, to put it mildly, from the customer. Long story short: we finally completed the work five months behind the original schedule and at a big loss to us. Instead of moving on with our lives, we consumed more energy and money filing a claim and going to arbitration.

Lessons Learned: Know what you are getting yourself into. Plan every aspect of your activities. Understand your costs and profit requirements. Manage the process and the players effectively. Don't panic—develop a strategy to deal with the variables that are beyond your control. Make sure that you have the proper company setup and controls to support your activities. Develop good communication and rapport with the customer from the moment you win a project.

The talents, skills, energy, and resources required to pull it off successfully can be overwhelming. The big question is whether you have what it takes. As the saying goes, "Luck happens when preparedness meets opportunity."

If you're serious about your business, you've got to start at the beginning with a comprehensive and realistic review of your skills, abilities, interests, financial situation, personal priorities, and goals. Taking inventory will help to confirm whether you have the skills and personality required to succeed and will enable you to identify the resources required to support your efforts. As you move ahead with your business, you can periodically refer to your self-assessment and make adjustments to ensure that you address your weaknesses and gather the resources you need to achieve your goals.

So, before you plunge into the meat of this book in Chapter 2 and beyond, please take stock of your disposition, capabilities, and resources. The Self-Assessment Questionnaire that follows provides a list of questions meant to facilitate this process. Sit down in a quiet corner and answer each of them honestly, checking off areas that you feel good about. The areas that you didn't check off you'll need to tackle head-on if you're serious about making a personal commitment to your business. Your answers to this questionnaire will also serve as the framework to help you think clearly, brainstorm intelligently, and move forward.

While writing this book, I interviewed many construction company owners and asked each of them this question, "What are the key ingredients to your success?" Here are some of their answers:

- I surround myself with "people in the know."
- I try to communicate a positive attitude and a clear focus at all times.
- I have good organizational skills.
- I'm a risk taker.
- I know I have shortcomings and rely on my advisors for support.
- I don't take a job unless I'm fully equipped to perform.
- I try to remain visible and responsive to my clients at all times.
- I evaluate each business opportunity carefully before jumping in.
- I keep a constant eye on costs, cash flow, and profitability.
- I don't nickel-and-dime anyone.
- I have strong relationships with my customers, subcontractors, and suppliers.
- I treat my personnel and customers the way I want to be treated.
- People love working for my company; we have good "chemistry."
- I have learned to be flexible and roll with the punches.
- I deliver what I promise.
- I know that trust, credibility, and loyalty are everything in this business.
- I can handle taking risks.
- My partner and I work well together.
- I have good relationships with my banker and bonding agent.
- I either do it right or don't do it at all.

SELF-ASSESSMENT QUESTIONNAIRE

Date: _____

- **The Right DNA.** It doesn't matter what business you're in, some character traits make it possible for you to get off the ground:
  - ❑ Do you have a positive "can do" attitude in life? Is your "glass half full"?
  - ❑ Are you self-confident?
  - ❑ Are you a leader?
  - ❑ Are you willing to work hard?
  - ❑ Are you a self-starter who can take the initiative and make decisions for others?
  - ❑ Are you an optimist with an ironclad will to succeed?
  - ❑ Do you have high energy, passion, drive, and determination?
  - ❑ Are you creative?
  - ❑ Are you adaptable and flexible?
  - ❑ Can you sell yourself and your business ideas?
  - ❑ Are you organized and self-disciplined?
  - ❑ Do you get along with people?
  - ❑ Are you able to look at "the big picture"?
  - ❑ Can you visualize your path?

- **A Strong Stomach.** To play the game, you need commitment, staying power, and guts:
  - ❑ Do you have the stomach to take risks and survive the ups and downs?
  - ❑ Can you handle drawn-out stress, problems, and uncertainty?
  - ❑ Can you hang in there and stay with it, despite obstacles or looming failure?
  - ❑ Do you have a tough skin that can protect you against rejection?
  - ❑ Can you wait patiently to see results?
  - ❑ Do you have clear-cut priorities?
  - ❑ Are you ready to make or renew your personal commitment?
  - ❑ Are you prepared to take full responsibility for your actions?

- **Knowledge and Expertise.** You must at least know what you don't know:
  - ❑ Do you have an educational background in a relevant field?
  - ❑ Do you have specific expertise and a track record in your line of business?
  - ❑ Are you informed about the current trends in the industry in general and in your field in particular?
  - ❑ Are you up-to-date on market conditions and opportunities?

SELF-ASSESSMENT QUESTIONNAIRE   **Date: _____**

- ❏ Do you have experience planning, coordinating, and managing all aspects of a project?
- ❏ Do you have a foundation of business knowledge and management skills?
- ❏ Do you have experience planning, organizing, and managing a business operation?
- ❏ Do you have experience in managing people?
- ❏ Are you comfortable with the basic principles of accounting/financial management?
- ❏ Are you up-to-date on information systems and software available for your business?
- ❏ Do you have a specific plan for the business? Can you visualize an organizational chart?

- **Financial Resources.** *You will not be able to start and sustain a business without access to money:*

  - ❏ If you're a start-up, do you have enough funds (savings, borrowings from a home equity line of credit, loans from family and friends) to cover your personal and business expenses until your company breaks even?
  - ❏ Have you made some projections about what types of expenses you will have and how much the annual budget will be this year?
  - ❏ Do you have partner(s) who are financially committed?
  - ❏ Do you have access to other investment capital?
  - ❏ Do you have a grip on managing the cash needs of your business?
  - ❏ Have you established a banking relationship? Do you have a line of credit? Is it sufficient to cover your anticipated needs?
  - ❏ Have you set up a user-friendly bookkeeping system?
  - ❏ Do you know where to go for help in these areas?

- **Support.** *No matter how good you are, you're always better off having a strong support system:*

  - ❏ Do you have partners or employees in your business who possess skills and experience that are important to your business and complement yours?
  - ❏ Are you aware of training programs and courses that would help you and your staff round out your collective skills?
  - ❏ Do you have industry and market information at your fingertips?
  - ❏ Do you have a lawyer, accountant, and bonding and insurance broker who are plugged into the construction industry? Do you keep them informed?
  - ❏ Do you have a mentor or advisor to talk to?
  - ❏ Do you have friends and colleagues with whom you can discuss industry and business issues?
  - ❏ Are you active in trade associations?

SAMPLE SELF-ASSESSMENT SUMMARY Date: _____

My Strengths:	Action to Be Taken:
1. Willing to work hard	Be careful not to burn out
2. Get along with people	Improve management skills
3. Am an optimist	Stay in touch with reality
4. Am organized and disciplined	Lead staff to do the same
5. Have a tough skin	Be sensitive to employees
6. Am a seasoned project manager	Stay current on means/methods
7. Have industry friends	Stay in touch for new business
8. Have a good partner	Discuss shareholders' agreement
9. Have vision/direction for the business	Confirm with market analysis

My Weaknesses:	Action to Be Taken:
1. Limited business experience	Take courses/seminars
2. No accounting and financial skills	Take courses/hire accountant
3. Don't know enough about systems	Take courses/hire consultant
4. Never written a business plan	Use workbook/get feedback
5. Low on capital	Talk to friends/banks for support
6. Not always up-to-date on what is going on	READ

Peter M. Lehrer is an internationally recognized industry expert who co-founded Lehrer McGovern in his kitchen in 1979. The construction company made its first mark restoring the Statue of Liberty and went on to become a top international construction company as Lehrer McGovern Bovis, with more than 1,700 employees and nearly $2 billion in revenues. I asked Lehrer to tell me what ingredients he believes are most important for success. Besides long hours of hard work, it's all about passion and focus. On passion, he said, "Passion should not be about money. It should be about vision. If you execute the vision well, the money will follow." About focus, he added, "Stay focused on what you know. What you don't know, don't do."

GETTING OUT OF YOUR OWN WAY

If you're confident that you're fully aware of the risks and challenges of your business and have the disposition, capabilities, and resources to be successful, then there's one last thing you must do before moving on to Chapter 2—get out of your own way!

Many construction company owners I've known over the years aren't ready to admit that they stand in the way of their own success. Their blind spots are curiously similar, and, hopefully, once you recognize them here, you will be sure to avoid them. I'd like to highlight a few:

- *Not recognizing the importance of strategic planning,* whether in relation to specific projects or to the entire business. Thus we see a frequent connection made between the terms *contractor* and *seat-of-the-pants management.* This is probably rooted in the fact that many contracting firms are family-run businesses headed by a patriarch who knows the craft but is weak on leadership and management skills. Another factor is the notion that planning is daunting and even irrelevant in an industry where no one is ever in full control of the variables.
- *Measuring success by looking at the top line (revenue) instead of the bottom line.* Most contractors are revenue-oriented; they're proud to say they are a $2 million, $5 million, or $10 million company. To gain volume, they accept low to nonexistent returns that don't compensate them for their risks, consequently lowering the bar for the entire industry. Chronically low profit margins are finally forcing many contractors to look at profitability, or "the bottom line," to gauge their progress and success.
- *Being "control freaks" about people and company management.* In their efforts to juggle commitments and responsibilities, company owners often fail to delegate properly and, instead, "micromanage" the business. Besides insufficient management experience, one reason may be a deep-seated lack of trust and confidence in employees and even partners. This creates a vicious cycle: CEOs are unable to develop strong teams capable of taking some or most of the burden off their shoulders and consequently end up managing the details of their businesses, often ineffectively.
- *Shying away from technology.* Contractors tend to take a long time to make a commitment to embrace systems and software that could help streamline operations, reduce costs, and improve communications with customers and project participants.
- *Not reaching out for help until the very last minute, if at all.* The causes of this are probably pride, fear of loss of privacy, and ego.

It takes a near miss or even a disaster before many contractors wake up and begin to change the way they do things. Please don't wait! Now is the time to do it right. We'll begin the process in Chapter 2 by looking at using top-down strategic thinking to formulate and maintain a workable plan for your business.

2

THINK "BIG PICTURE"

recently called a client to arrange a meeting to discuss strategic planning for her business. Her response was, "I'm not ready yet. I've been too busy putting out fires." She delays thinking about overall goals and finds herself in a permanent crisis management mode. Business owners of every stripe avoid business planning. It can be time-consuming and tedious and diverts attention from day-to-day business activities.

Like my client, you've heard umpteen times about the importance of strategic planning but may not have gotten around to doing it yet. There's always next week or even the weekend to revisit your original goals, where you are now and where you are heading. You may have it all sorted out in your mind and wish to download it on paper and perhaps share it with your team, but, first, several pressing matters must be settled before week's end.

It's understandable; daily pressures keep you from looking at the big picture. Your daily challenge is to bid and win as much work as possible to keep your operation going and make a profit. Many of you wear various hats throughout the day, from CEO to CFO to project manager to site superintendent and administrative assistant. You're so busy chasing after new business and attending to current work that you put off the brainstorming required to figure out what you should be bidding on in the first place!

PLANNING TO PLAN

Whether yours is a start-up or an established business engaged in residential, commercial, or industrial construction—as a general contractor, construction manager, or any type of specialty contractor—you must not allow this vicious cycle to go on indefinitely.

And it's never too late. I know of one general contractor who has finally decided to put together a business plan in his 27th year in business! His sense of urgency now comes from his desire to wring out the weaknesses that still exist, review the company's position in the current market, and set the company on the right course for the next five years.

Everything you do flows from one strategic question: Where do you want to go with the business? Without a game plan, you'll fall into the trap that most contractors find themselves in—chasing after every project that comes along and leaving it to the luck of the draw. You can't afford to fritter away your energies and resources "barking up the wrong tree." Winning a job that doesn't suit your business is a recipe for trouble and perhaps even disaster. Commenting on this topic, industry veteran Peter Lehrer told me, "In many cases, saying no is the right answer."

In addition, you face unique industry challenges on a daily basis. To *insulate* your business from those outside forces as much as possible, it's imperative that your company organization be *at least* adequate to support the level of business you're doing today and anticipate for tomorrow. To accomplish this, you must create a workable business structure that can adapt to change within an overall framework. The business plan is that framework.

Planning is a process of exploring yourself and your business that will help you to:

- organize your thoughts and gain perspective on the realities of the construction industry;

- encourage you to analyze, evaluate, prioritize, and articulate your goals and expectations;
- establish and maintain long-range focus;
- recognize and deal with your limitations;
- minimize panic situations and 24/7 preoccupation with daily problems; and
- become a better manager.

Put in writing, your plan can then evolve into an essential management tool to help establish performance benchmarks, track progress, and encourage active participation and commitment from your team. It's meant to be reviewed and refined periodically as the situation changes (e.g., changes in economic and business environments, costs, competition, technology, government regulations).

As a formal document, the business plan will also come in handy as an effective and persuasive communication tool; it will go a long way toward convincing potential bankers, investors, and bonding companies to provide financing, capital, and support for your business.

Above all, the exercise puts you in the driver's seat. It'll help you feel confident about your business potential and give you the courage to carry on. When you're disheartened, you can refer back to your plan and remind yourself that you are, indeed, on the right track!

Planning isn't something you can hand off to a professional business plan writer or plug into an off-the-shelf business plan software package. You've got to do the hard thinking yourself and then commit your goals, strategies, and action plans to paper. You probably just need a push and some guidance to get you started.

DOING IT NOW

You're probably saying to yourself right now, "Hold on! I haven't got the time or inclination to sit down and write a 50-page thesis!" There's no need for you to do that. Let's go through the steps you can take to put your plans on paper as painlessly as possible.

In Chapter 1, you identified your personal strengths and limitations and the resources required to round out your capabilities. In this chapter, you will use your Self-Assessment Questionnaire as a basis for moving forward with your plan. I'll take you through five important steps:

1. Outline your personal and business goals, the current obstacles, and the resources required.
2. Set the parameters that define your business.
3. Take a look at market conditions and trends and confirm that your goals are achievable.
4. Prepare a business plan that you can expand on and formalize later. At this stage, use phrases or bullet points to make it easier for you to jot down ideas and plans.
5. Determine how you will use the results of this exercise as a planning and management tool going forward.

I will provide worksheets in the chapter for you to pencil in your initial thoughts. The Goals Worksheet in this section will help you set goals and priorities and the Strategic Business Plan Workbook (see Appendix A) will provide a simple question-and-answer format to follow. You can download a digital copy by going to *www.ConstructBiz.com*.

Once you've completed your Strategic Business Plan Workbook, I suggest you obtain feedback and suggestions from knowledgeable sources such as your accountant, bonding agent, or business advisor. For free counseling, you can reach out to a local chapter of the Service Corps of Retired Executives (SCORE) to enlist the help of a volunteer retired business owner. You can also go to a Small Business Development Center (SBDC) at a local college for workshops, classes, and one-on-one counseling. Both programs are sponsored by the U.S. Small Business Administration. The SBA also offers tips on business plan preparation and a sample "Business Plan for Small Construction Firms" that is accessible via its Web site (*www.sba.gov/library/pubs/mp-5*). In addition, reach out to your local chamber of commerce and trade associations such as the Associated General Contractors of America (AGC) and the American Subcontractors Association (ASA).

Step #1: Confirm Your Personal and Business Goals

In a private business, it's difficult if not impossible to separate personal from business goals. As a business owner, you must achieve a degree of personal satisfaction, fulfillment, independence, and financial stability that make it all worthwhile, or you'll be miserable. That's why you decided to take the leap in the first place! You also want to ensure that the business is working for you and not the other way around.

Your personal goals are achievable only if specific, realistic business goals and targets are met. Thus, your personal and business goals must be compatible and mutually supportive and reflect your overall philosophy and aspirations. If not, you may suffer from *cognitive dissonance* (a nagging feeling that things aren't right) and lose sight of the success and peace of mind you seek. Ideally, you'll have a personal vision and a shared mission statement, the latter being the vision for your company that defines it for your team, customers, suppliers, and the world at large.

I've designed the Goals Worksheet that follows to assist you. Let's break down both personal and business goals into manageable steps by thinking about your overall goals first. With your goals in mind, you can formulate long-term objectives (five years and beyond) and, from there, identify short-term objectives (one year). This exercise should be done periodically (for example, quarterly) so that it's always fresh and current and helps focus your attention where it really counts.

Then, identify specific obstacles that are stopping you from getting there, both external conditions and internal factors (e.g., staff capabilities, availability of capital). Lastly, review the resources and skills you currently have to determine where gaps must be filled and how. Be realistic—painting a rosy picture for yourself gets you nowhere!

Step #2: Define Your Business

In the Goals Worksheet, you paint a "big picture" view of your business in broad brushstrokes by defining your overall vision and goals, and the hurdles that you must overcome to be successful. This analysis forms the foundation for your next task, which is to define the characteristics of your business. Digging back into your Self-Assessment Questionnaire in Chapter 1, be sure that your business description is realistic, given your experience, capabilities, and resources.

GOALS WORKSHEET   Date: _____

Overall Vision

- One sentence for personal goal:

Business

Example

To build a company that "runs itself" so I can reap the rewards and enjoy my life.

- One sentence for mission statement:

To develop an organization that delivers top-quality performance and service to our customer base.

Long-Term Goals

PERSONAL

1.

2.

3.

4.

5.

Work on big picture planning.

Develop people who can run the company.

Reduce time spent in the business.

Remove personal guarantees from loans.

Sell the company in eight to ten years.

BUSINESS

1.

2.

3.

4.

5.

Develop a reputation for quality service.

Build a solid organization.

Enter new markets.

Improve profitability.

Develop company to be valuable asset.

Short-Term Goals

PERSONAL

1.

2.

3.

4.

5.

Spend more time doing marketing.

Take management workshop/seminar.

Set up retirement plan.

Give myself a raise.

Have more fun.

BUSINESS

1.   *Win three or four new projects.*
2.   *Finish business plan.*
3.   *Begin next year's marketing plan.*
4.   *Hire new project personnel.*
5.   *Reduce expenses for net profit before tax of 10 percent.*

Obstacles   **Remedies/Action Steps**

INTERNAL

1.
2.
3.
4.

EXTERNAL

1.
2.
3.
4.

Resources Required

EXTERNAL

1.
2.
3.
4.

INTERNAL

1.
2.
3.
4.

DESCRIPTION OF YOUR BUSINESS

Date: _____

Key Characteristics

Comments

1. Type of Business:

Trade/specialty contractor, general contractor, construction manager, design/builder

These businesses are in many ways vastly different, requiring specific expertise, management know-how, and business philosophy.

2. Type of Work:

Residential, commercial, institutional, public sector, private sector

You could be involved in one or more of these sectors based on your experience and goals.

3. Geographic Span:

Local, regional, national

Make sure that you have the proper licenses in your geographic area.

4. Specific Areas of Expertise:

Office, retail, schools, hospitals, government buildings, private homes, etc.

These are based on your team's history of experience and your goals.

5. Legal Structure

Sole proprietorship, partnership, corporation

Confirm with your attorney that your legal structure is best suited to your business and that it protects you from personal risk.

6. Small Business and Minority Status:

Small, disadvantaged, minority, or woman-owned (SBE, DBE, MBE, or WBE)

There are many advantages to becoming certified in your state, especially if you work in the public sector.

7. Affiliation to Organized Labor:

Union or nonunion shop

In some regions, it may be advantageous to be a union shop to obtain skilled labor.

Step #3: Define Your Market

Whether you're just starting out or have been in business for many years, you've got to keep your finger on the pulse of the industry at all times. You must understand the forces at play to conduct yourself properly in your particular environment.

This is even more vital today than it was ten years ago, because the information age has dramatically altered how we do business and has accelerated exponentially the pace of change. Fortunately, virtually all the information you need is at your fingertips if you know how to find it. The more you know about your market, the clearer your path becomes and the more confident you are that your personal and business goals are achievable.

The big challenge is to block time out of your overscheduled days to read, research, and network. After years of struggling with this, I've decided to accumulate periodicals, articles, and other information during the week and read them at my leisure in transit or on weekends. The result is that my desk is neat and I'm a lot more focused at the office! I delegate the legwork once I've identified key areas for further research. If you provide specific instructions, someone in your staff can go online and obtain the raw information you need to analyze a situation. If you don't have someone to help you, try to devote half an hour every day to "surf the net" and research one or two topics on your list.

Here are examples of where you can go for the latest information:

- General business magazines, newspapers, and newsletters (e.g., *BusinessWeek, Crain's, Entrepreneur, USA Today, The New York Times, Kiplinger's*)
- National and regional trade periodicals (e.g., *McGraw-Hill's Engineering News-Record, Architectural Record and Construction News; Real Estate Journal; Building Design & Construction, Contractor, Construction Business Owner*)
- Business listings such as the *Blue Book* and *Dun & Bradstreet*
- Listings of project opportunities and industry information provided by the *F.W. Dodge Reports, Reed Construction Data, BidClerk.com*, and others
- Trade associations for specific industry statistics and information and related publications

- Trade shows and industry conferences
- Government agency programs, project listings, and resources
- Industry hubs such as *www.construction.com* and its McGraw-Hill newsletter and linked resources; Construction Data Company's *www.cdcnews.com; Professional Builder's www.probuilder.com* and search center, *www.info.com/construction*
- Online book sellers such as *www.constructionbook.com* and *www.bookmarki.com*
- Competitor Web sites

You can go to the appendixes in the back of the book to obtain detailed online listings of: industry resources and information (Appendix B); government resources (Appendix C); minority and women business organizations (Appendix D); industry associations (Appendix E); and construction unions (Appendix F).

Your most direct source of information, of course, is the people you deal with every day. Make a real effort to establish and maintain a dialogue with your employees, industry colleagues and friends, subcontractors and suppliers, accountant, attorney, banker, insurance broker, and surety agent. Discuss your issues and concerns, ask questions, and exchange ideas and information. Join and actively participate in trade associations that are relevant to your business, and network with industry colleagues, subcontractors, and even competitors.

As a business owner, your eye must always be on the *external conditions* that affect your business. You must keep track of overall economic factors that impact the industry, areas of growth opportunity and weakness, key players, regulatory changes impacting construction, the latest in project delivery methods, sources of supply and procurement methods, and project management practices and systems. You also should be plugged into developments in the worlds of financing, insurance, bonding, and contract law.

Most important, keep a close watch on conditions in your local market. What are the political and economic factors that impact your specific geographic market? Where are specific areas of opportunity for your business? Who are your immediate competitors and how do you compare? What does the local pool of subcontractors and suppliers look like? What are the trends in costs and the availability of labor, materials, equipment, and supplies? What is the current labor environment? What banks and sureties are active in your market and what do they offer? Where can you go for help?

Your legwork should lead to an understanding of industry and market opportunities and risks, current and potential markets for your business, characteristics of your target prospects, strengths and weaknesses of your competitors, sources of supply, potential niches for your business, and the resources you will need to compete effectively.

Armed with this information, you can put your business ideas into perspective and plan your goals and approach accordingly. At any given time, you should be able to respond to these essential questions:

- What business am I in?
- What is my specific market?
- What do I have to offer?
- Who is my competition?
- How can I compete effectively?
- What company resources do I need to operate successfully?
- How much money do I need to operate the business?
- What are my strengths, weaknesses, opportunities, and threats?
- What is the *key* determinant of my success?

Step #4: Design a Workable Strategic Plan

Ultimately, your objective is to have a realistic action plan with specific milestones and deadlines. Working with the results of your Goals Worksheet, now use the Strategic Business Plan Workbook located in Appendix A to get started on your business plan. The workbook is in outline format with questions you can answer using simple phrases and bullet points. Don't be concerned about how it reads just yet—at this stage your work is for your eyes only. The workbook is divided into the following sections:

- Executive Summary
- Business Description
- Products and Services
- Markets and Competition
- Marketing
- Company Operations

- Management Team
- Financial History and Requirements
- Appendixes (backup marketing, project, client, and financial information)

Two alternative suggestions on how you can tackle this exercise are:

1. Take time out now to complete the workbook as much as you can and finalize it as you move along in the book.
2. Work on each section while you're reading the related topic in the book. Either way, you'll be prompted to respond to the following key questions.

Executive Summary

The Executive Summary, a synopsis of what your company is all about, can be polished to become a concise, powerful, and persuasive sales pitch to be shared with the world. Leave this section for last, after you have completed the entire workbook.

Business Description

- What business are you in (general contractor, plumbing contractor, asbestos removal contractor, etc.)? How long have you been in business? Who are the owners?
- How do you describe your track record to date? From the Goals Worksheet, what are the biggest obstacles facing you today? What resources do you need?
- What is your company's legal structure (sole proprietorship, partnership, or corporation—C, S, LLC, LLP)? Do you have any joint ventures or formal alliances? How about partnership or shareholder agreements?
- Are you licensed to do business in your geographic area? If you are a minority-owned or woman-owned business, do you plan to obtain government certification so you can participate in special procurement programs?
- From the Goals Worksheet, what is the mission of your company? Where do you ultimately want to take the business? What are your top five goals for the next five years? What are some of your specific objectives relative to revenues, profitability, new business, company growth, and market presence in the next year or two?

Products and Services

- What specific services do you provide?
- What value do you bring to your customers? What problems do you solve?
- Do you have any proprietary products (with patents, trademarks, or copyrights)?
- How is your service unique or different?
- What level and types of customer service do you provide?

Markets and Competition

- What are the current trends in the market? What factors most impact your business?
- How do you define your target market (geographic market, customer profile, project type, project size)?
- Do you have an established market niche?
- Who are your competitors? Which ones present your biggest threat?
- On what basis do you compete (e.g., pricing, expertise, quality, reliability, service, successful project track record)?

Marketing

- Have you identified potential growth areas for your business? What services should be scaled back or de-emphasized? What new services should you develop?
- How have you positioned your company in your immediate market? What advantages do you offer? What are you best known for?
- Who are your customers? What customer groups do you want to reach?
- How are you marketing your company? Do you have a professional-looking brochure, business card and letterhead, a Web site, and other marketing materials?
- How much money are you spending on advertising, networking, and marketing? What are the results so far? What is your image?
- What are your estimating and bidding capabilities? Do you have a good grasp of your costs and how you should price your services in the market?
- What is your approach to servicing customer needs to ensure repeat business?

Company Operations

- Where is your business located? What are the benefits and drawbacks to your location?

- What equipment and facilities do you currently have? How much will it cost to purchase new equipment? Do you own or lease?

- What does your organization chart look like right now? What gaps need to be filled in the short and medium term?

- What are your personnel requirements? How many workers do you need? Will you use union or nonunion labor? Who will supervise them? Who will manage the office?

- How do you go about hiring employees? Do you offer training and a career path? Have you written job descriptions for each position? How do you evaluate your employees' performance? Do you have a personnel manual? What job functions could you outsource?

- Do you have a strategy and methodology for estimating, bidding, and buying out projects? How do you manage project setup and staffing?

- Do you have established practices for managing a project? How do you control project quality, costs, and schedule?

- Do you have manuals, forms, technical reference books, and software available in the office?

- How do you manage procurement? Do you prequalify subcontractors and vendors? How do you describe your relationships with them?

- What bonding capability and insurance coverage do you have? What do you anticipate your needs to be next year? What is your safety record? Do you have any pending legal claims or litigation?

Management Team

- Who manages the company? Who are your key office and field personnel?

- What are the credentials, experience, and responsibilities of your key team members?

- Do you have personnel gaps that need to be filled?

- Who do you rely on for outside expertise, support, or as a sounding board?

- If something happens to you or one of your partners, who takes over? Do you have a shareholder buy-sell agreement? Do you have a business continuity plan?

Financial History and Requirements

- What is your company's financial history to date (revenues, profitability, balance sheet strength)?
- If you are a start-up, how much money do you need to cover personal and business expenses until the company can sustain itself? How much money can you raise between your savings, home equity loan, loans from family and friends, and investments from partners?
- Do you have credit lines? How often and how much do you borrow? What are your needs for the foreseeable future? How do you intend to repay the borrowed funds?
- Do you have access to bonding? How do you plan to obtain more if or when you need it?
- Do you have outside investors? What are their expectations? How will you pay them out?
- Do you know what your break-even point is? How do you manage your costs?
- Do you have a grip on the day-to-day cash flow needs of the business and for individual projects?
- How do you manage billing and collections?
- What are your goals for revenue growth and profits for the next year, the next five years?
- Based on current business backlog and assumptions regarding new business, what are your financial projections for this year and the next?

Appendixes

- Financial statements
- Financial projections
- Work in progress and project backlog information
- Customer references
- Résumés
- Personal financial statements and guarantees
- Construction contracts

- Corporate legal documents
- Insurance policies, leases

Step #5: Use the Plan As a Management and Financing Tool

Provide Leadership and Manage the Big Picture

Even if, at this stage, your business plan is only a handwritten draft with incomplete sentences and grammatical errors, it starts working for you right away. It empowers you to manage the future direction of your company and handle day-to-day issues as they arise.

By understanding the external forces affecting your business and establishing goals that make sense within that context, you can act quickly to address opportunities and solve problems. You'll be able to prioritize your goals and objectives and identify the resources you need to achieve them. You'll possess the wherewithal to take advantage of your company's strengths and work to reduce weaknesses and threats. And it will become clear how you should organize your company and delegate responsibilities to meet your objectives.

Create Team Enthusiasm and Commitment

Once you've established your major goals and objectives, you can enlist the help of your staff to formulate specific strategies and solutions. Key team members can be encouraged to generate ideas and solutions in their areas of expertise and help set short-term objectives. You can promote efficiency and teamwork by assigning responsibility and target dates to each participant to implement action plans and report on progress. Such a group effort leads to team building, creates a sense of commitment and participation, and leads to opportunities for effective brainstorming and problem resolution.

Review Progress, Measure Results

Once you've set long-term goals, short-term objectives, and immediate action plans, it's crucial that you set aside a block of time every month, perhaps away from the office, to review progress. Review your business activity, revenues, costs, profitability, working capital, and other financial results compared to your benchmarks and flag areas that need your attention. Meet regularly with your key team members to review progress on specific action plans and keep the momentum going. Be prepared to make adjustments to your plans in light of experience or changed circumstances. If Plan A no longer works, have Plan B ready to go.

Obtain Financing and Bonding

Once you finalize and polish your business plan, it can be used as a tool to convince your banker to provide you with the financing you need. As an ex-banker, I know firsthand that most lenders don't immediately get a "warm and fuzzy feeling" when you walk in the door because they don't understand the nature of your business and they possess preconceived notions about the construction industry. What they do know is that construction is a risky business with high rates of business failure and that you probably don't have a lot of business assets with which to secure a loan. By sharing your business plan with your banker, you will dispell this negative view and open up a positive dialogue with him or her.

Your plan will also be a big plus when you try to obtain bonding. Surety companies will want to see a business plan and a management succession plan. They'll be particularly concerned with business continuity if something happens to you or another key person.

A simple, concise, well-organized business plan (bullet-point format is fine) will go a long way toward helping your banker and bonding agent feel comfortable. It will show him or her that you are in control of your business; that you know what your strengths and weaknesses are; that you have a realistic game plan and know what resources you need to get there. It will also demonstrate that you are developing a solid organization that can function without you if need be.

First impressions are critical, so make sure that your Executive Summary delivers the punch line; it must clearly describe your business and track record, major opportunities and threats, overall goals and objectives, the purpose of your request for financing, and how you intend to repay the loan. Provide only the facts; giving a banker fluff and an overly rosy picture will surely backfire. As one prominent construction lender said to me recently, "*Never* overpromise. We want to see that you always exceed your expectations!"

You'll have to corroborate your request in the body of the business plan by providing important facts, solid analysis, realistic best/worst/ most likely case scenarios and contingency plans. You'll provide no-nonsense numbers that are backed by reasonable assumptions. You'll also confirm that you have the right team, company organization, and resources to achieve your goals.

Before approaching the bank or bonding company, I suggest that you show your plan to someone who is knowledgeable about business and can provide you with constructive feedback. Your accountant could be a great help in putting together financial information and developing financial projections.

More Reading on Business Planning

Banks, David H. *The Business Planning Guide,* 9th Ed. Chicago: Kaplan Publishing, 2002.

Harris, Jr., H. M. *Strategic Planning for Contractors.* FMI Corporation, 2001.

Business Plan for Small Construction Firms. Management Planning Series, U.S. Small Business Administration *(www.sba.gov/library/pubs/mp-5).*

3

CREATE A REALISTIC MARKETING PLAN

As a contractor, your main objective is to bid and win enough projects throughout the year to keep your resources deployed and make a profit. To accomplish this, you must win a healthy backlog of work and create a pipeline of viable business opportunities. If not, you run the risk of standing at the edge of the cliff every so often, panic-stricken, worrying whether you'll ever work again.

Truth be told, marketing has not been on the top ten priority list for many a construction contractor. Who has time for it? You are consumed with finishing existing work and bidding on the jobs that come your way. And why bother? You wonder whether there's a reason to create a detailed strategy and glossy marketing materials in an industry where the key determinant to winning new business is price, and where, in the public sector, project opportunities are publicly announced and bidders galore show up at the prebid meeting.

In the private sector, price may be weighed along with subjective criteria such as track record, reliability, quality of the work and service, integrity, professionalism, and even personal rapport. So, some degree of marketing and networking is necessary to get in the door with a prospect. Minimum requirements include professional marketing materials and a referral or two to facilitate an introduction.

Once in the door, however, contractors tend to rely on customers to "invite" them back to bid on the next job and to refer them to other opportunities. A contractor can survive and even thrive for years relying on repeat business from a couple of customers.

That sounds great. The only problem is that efforts to market to a wider universe tend to wind up on the back burner. One day, the contractor loses an important customer—who went out of business, left the state, ran out of money, refocused strategy, shifted resources, found another preferred contractor—and, *poof,* revenues drop suddenly. The contractor then scrambles to replace the lost business hoping that, by some miracle, something turns up in time to save the day.

Watch out, because the world has become a lot tougher lately. As mentioned in Chapter 1, projects have become increasingly complex because of various factors, including tight budgets and schedules, stricter safety and environmental requirements, new building technologies, shortages of qualified personnel, and fluctuating materials prices. Public- and private-sector owners have become more demanding and discerning, requiring contractors to demonstrate relevant experience, a solid track record, and a strong management team that has *specific expertise* in the customer's type of business. Hank Harris, management consultant to contractors, warns that "Your competitive edge comes from understanding your customer's business."

As one contractor aptly put it, "Unless you're a megacontractor, the market will not come to you." So now, more than ever, you must appreciate the critical role marketing plays in business development. Professional marketing materials and a marketing plan are musts if you intend to create the steady flow of business required to succeed.

Some CEOs I know are trying to do it all on their own while others are hiring or outsourcing sales and marketing expertise. Either way, your efforts won't yield positive results unless there's a real commitment to a consistent strategy and a willingness to invest a reasonable amount of effort and money.

POSITIONING YOUR COMPANY

If you look up your particular business segment (e.g., carpentry, masonry, general contracting) in the Yellow Pages or in an industry listing like the Blue Book, you'll see hundreds if not thousands of company names listed in fine print along with yours. If you were to look at random samples of their brochures or Web sites, you'd see that they all promise to deliver quality work on time and on budget.

You'll be reminded that you're in an industry with low barriers to entry, where you're "one of the pack." If all of you provide the same basic service, how can you possibly distinguish your company enough to stand out? How can you communicate that you are unique, different, better? How can you *position* yourself so you will be noticed and eventually selected?

In Chapter 2, you began jotting down your personal and business goals and objectives and hopefully started brainstorming on strategies and action plans that will enable you to meet your expectations. Here, you'll focus on your core competencies and services and create a Market Positioning Statement.

I recommend that you first do some homework:

1. *Make a list of your most immediate competitors* and the largest players in your market and visit their Web sites. What are their primary areas of work and how many employees do they have? What are their strengths and weaknesses? What are the credentials of the principals? Do they have company logos, mottoes, or mission statements? Do they send clear and believable messages? List your findings on one spreadsheet so you can look at a complete picture and start thinking about the content of your own message.

2. *Obtain feedback from your customers.* How do they go about making a list of contractors? What selection criteria do they use? What are their priorities? What services do they need that they're not getting? How satisfied are they with your services? What areas could be improved? You can design a simple one-page questionnaire and send it to your customers as projects are completed. You can also meet with your biggest customers in person and discuss your performance, their concerns and future needs. They'll undoubtedly appreciate your interest in their candid feedback!

Armed with all this information, and using the template that follows, create your Market Positioning Statement. The statement's purpose is to summarize who you are, what you believe in, what you do best, what benefits you offer, and how you want to be perceived in the market. You'll choose words and phrases that will be the basis for a powerful and consistent message to your target audience. The exercise will also help you focus your energy and resources on creating a target market or niche and help you establish your market identity.

IDENTIFYING YOUR TARGET CUSTOMERS

By completing the Market Positioning Statement, you've defined the business you're in, the markets you serve, the benefits you provide, and the experience and expertise you bring to the table. Armed with this basic information, and your knowledge of market conditions and areas of growth and weakness, you can now choose the types of customers you intend to do business with. By preparing a Target List and updating it annually, you will be able to create and implement an effective marketing strategy consistent with your overall business goals.

Before you begin to create your Target List, I'd like to caution you on three fronts:

1. Your top priority should be to maximize the amount of *repeat business* you generate, which means making a concerted effort to maintain strong relationships with customers who are active. If you're a trade contractor, it's *critical* that you develop good relationships with general contractors so they'll invite you to bid on their projects. It's very expensive, time-consuming, and frustrating to create a steady pipeline of business from prospects only.

2. Look to achieve *controlled growth*, meaning that you choose reasonable targets for your company in terms of scope of work, size, and type of project given your experience and resources. Contractors risk losing it all when they overextend themselves to work on larger or more complex projects than they are accustomed to, or if they move in an entirely different direction or to another location.

3. Take a *balanced approach*. Instead of chasing after what is currently "hot" in the market or what you've grown comfortable with, try to spread your risk and look at different markets where

MARKET POSITIONING STATEMENT   Date: _____

Your Statement

Example

1. What business am I in?

General contractor serving private-sector building owners and developers

2. What services do I provide?

Lump-sum and negotiated work, including new construction, renovations, interior fit-outs, and historic restorations

3. Who is(are) my target market(s)?

Retailers
Entertainment companies
Corporate and financial
Cultural institutions, private schools

4. What is my advantage?

Active principal involvement
Sensitivity to customer needs
Excellent service and follow-up
Top-notch management and field team
Reliable subcontractors

5. What are my credentials and track record?

Twenty years of industry experience
Proven ten-year track record
High customer satisfaction and repeat business
Brand-name customers

6. WHAT IS MY MESSAGE?

We jump through hoops for our customers.

your work can be performed. One suggestion would be to think about widening your customer base to include new types of customers. Another would be to spread your work between the public and private sectors; it's often true that when the private sector is weak, the public sector is strong, and vice versa. By diversifying your client base, you can help protect your company from the waves, shocks, and bumps that characterize the cyclical nature of the construction industry.

DEVELOPING YOUR PLAN

Based on your Market Positioning Plan and Target List, you can develop a marketing plan that is uniquely suited to your goals and objectives and the attributes of your company.

Your plan will be grounded in your knowledge of the business and your intuition. Who else knows your business as well as you? I would suggest, however, that you review it with someone who is knowledgeable in marketing (a friend, colleague, or consultant) before you finalize and implement it. Why? You're not a Fortune 500 company that can spend millions experimenting with different approaches. With limited resources, you have no choice but to commit to your plan and stay the course for a period of time so you get the biggest bang for your buck. There's another reason why you should stay with your plan: people forget fast, even customers. You want to maintain a presence and an identity in a market where "out of sight" is definitely "out of mind."

And don't expect instant results; it takes time for your marketing efforts to start paying off. So if you panic and change course every few months or suddenly drop your plan, you will have little to show for your efforts besides higher "marketing and promotion" expenses in your income statement.

I recommend that your plan be brief and focused—a few paragraphs or bullet points—instead of a lengthy and detailed document that you will dread reading later. Following is a Summary Marketing Strategy worksheet to help you get started.

TARGET LIST

Date: _____

Experience	Targets

Experience

1. Project size

$_____ to $_____

Targets

$_____ to $_____

2. Geographic area(s)

—

3. Areas of expertise

(e.g., plumbing, general contracting, painting, masonry, roofing)

—

—

4. Type of work

(e.g., renovation, new construction, historic restoration, rehabilitation, highrise construction)

—

—

—

5. Type of end user

(e.g., residential, educational, retail, warehouses, corporate offices, health care, commercial, roadwork, transportation, industrial, site work, landscaping)

—

—

—

—

6. Private-sector customers

(e.g., for trade contractor: prime contractors; for general contractor: building owners/developers/tenants, construction managers, design/builders)

—

—

—

—

7. Public-sector customers (federal, state, local, quasi-public, utilities)

(e.g., for trade contractor: prime contractors; for general contractor: government agencies, construction managers, design/builders)

—

—

—

—

8. Primary contacts

(e.g., for trade contractor: prime contractors; for general contractor: owners, architects, owner representatives, construction managers)

—

—

—

—

9. Secondary contacts

(e.g., real estate brokers, vendors, suppliers)

—

—

10. Tertiary contacts

(e.g., trade association members, friends)

—

—

SUMMARY MARKETING STRATEGY Date: _____

Overall Objectives:

Example

-
-
-
-
-

Expand customer base (three or four new)
Identify a new "anchor" customer
Balance public- and private-sector work
Explore strategic alliances
Improve/update marketing materials

Services Offerings:

-
-
-
-
-

Add project planning services
Add postproject maintenance services
Improve customer follow-up

Existing Customer Base:

-
-
-
-
-

Seek repeat business from customers
Obtain referrals and testimonials
Network with associated architects, contractors, consultants

Target Customers:

-
-
-
-
-

Highrise residential
Private schools
High-end retail rollouts
Multiplex movie theater chains
Public schools and housing

Target Contacts:

-
-
-
-
-

Real estate brokers
Architects
General contractors (if sub)
Building owners and developers
Government agencies

Media Plan:
- Advertise in trade journals
- Submit press releases to local press
- Write opinion or expert articles
- Renew company listings

Networking:
- Prepare sales presentations and materials
- Maintain membership in trade associations
- Sponsor an industry association meeting
- Attend government agency conferences
- Meet with contacts regularly
- Host a networking event
- Create a trade show exhibit

Packaging:
- Update brochure
- Design postcards, newsletters
- Plan regular promotional mailings
- Design company site sign

Web Site:
- Upgrade and expand Web site
- Link to industry Web sites, search engines
- Create user-friendly "customer" or "subcontractor" areas

Strategic Alliances
- Identify joint-venture opportunities
- Explore strategic alliances with vendors and suppliers
-

Other:
- Explore new office location
- Search for part-time marketing person
- Consider hiring a PR firm

Marketing to the Public Sector

Most government agencies that undertake or sponsor construction projects are required to publicly advertise a bid opportunity and accept all bids for consideration. Unless a problem is uncovered after bid opening (e.g., the winning bidder doesn't have the required bonding or insurance), the "lowest responsive bidder" wins the job.

When a contractor is chosen based primarily on the bid price and a bond, the customer is not necessarily getting the most qualified company to do the job. The contractor may have lowballed the bid to win the job and is counting on "change orders" (extra work) to make up the difference. Or, worse yet, perhaps the contractor made a mistake in the bid by omitting an important item, making mathematical errors or miscalculating certain costs.

Concerned with poor contractor performance and creeping project costs, many public-sector agencies are trying to raise the bar on the "lowest bidder" concept by requiring contractors to undergo a prequalification process before they can bid on work.

Several public agencies have figured out ways to register and prequalify contractors in efforts to improve the quality of the bids. Instead of advertising notices to bid, they first put out a "request for qualifications" and thereby create a list of qualified bidders. Those who meet the requirements will be allowed to bid on the project and will be selected based on price.

This means that if you want to work for certain public agencies in your market area, you must take a proactive approach instead of waiting for bid announcements:

1. *Make a list of the agencies* you believe need your services and study their Web sites to learn about their policies, procedures, and requirements regarding bids.
2. *Register your company* with the agency so it has your company contact information on file.
3. *Keep track of project bidding opportunities* and "Requests for Qualifications" that are listed on their Web sites or announced publicly. Make sure you're on the mailing list (and/or e-mail list) to receive information.
4. *Consult government agency Web sites* to confirm which standard forms must be completed. Several forms (i.e., the S.F. 254 and 255

Sample Outline for Contractor Qualifications Statement

I. Letter of Interest

Stating why your company should be prequalified for the project

2. Qualifications Statement

a. Company History

Years in business; company organization; types of construction; track record; typical project size; current workload; safety record

b. Project Experience and References

Minimum three projects of similar scope, size, and design quality; project name, location and description; construction budget; completion date; owner name and contact information

c. Financial Capability

Bonding capacity; credit lines; insurance coverage

Financial statements for past two years

d. Project Organization/Staffing

Staffing plan; project organization chart, résumés of project personnel

and the S.F. 330 forms, the Uniform Contractor's Questionnaire) are used widely by federal, state, and even municipal agencies. Download these forms, complete them correctly, and keep them on digital file so you can update them at any time for inclusion in a prequalification or bid package. This saves you last-minute aggravation.

5. *Create basic marketing materials* that include the information in the Sample Outline for Contractor Qualifications Statement provided here. Your materials should be up-to-date, well organized, comprehensive, responsive, and professional. If necessary, seek help from someone experienced in putting these packages together and ask someone in your office to proofread everything. No typos, please!

6. *Organize your financial and insurance information.* Make sure that your accountant prepares annual financial statements and that your bonding company is ready to confirm bonding capacity. If you have a line of credit, ask your banker to provide a letter confirming the amount of the line and that you are in good standing.

Also, confirm that your insurance coverage is current and meets the agency's minimum requirements.

7. *Attend conferences* held or sponsored by the agencies you'd like to work with. You'll learn about current and planned project activities and overall agency construction budgets. You'll also have a unique opportunity to introduce yourself in person to agency staff, prime contractors, and consultants with whom you'd like to work as a subcontractor or as a supplier.

8. *If you are a trade contractor, obtain information about prime contractors* who do work for the agency. Contact them directly and send them your company information.

9. *Join affirmative-action programs.* If you're a minority-owned, disadvantaged, or woman-owned business, research federal and state public agency outreach programs. Make it a priority to certify your company in your state or other relevant jurisdiction and learn about special mentoring and "set-aside" programs that are offered specifically for minority-owned, disadvantaged, small, and women-owned companies (MBEs, DBEs, SBEs, and WBEs). You'll find the information you need on public agency Web sites.

Marketing to the Private Sector

Private-sector end users are looking for track records, financial capability, management talent, work quality, integrity, professionalism, responsiveness, and cooperation. They're looking for contractors they can *trust* and *rely on* to take care of their needs as they arise. If you can demonstrate that you meet these criteria and you are chosen, your bid price will most likely be *negotiated.* Your reward is the promise of word-of-mouth advertising, referrals, and repeat business.

What are private-sector project owners looking for in a contractor? I asked that question of a variety of project owners with ongoing construction programs. Marcelo Velez, director of design and construction at Columbia University, where he has presided over $1 billion in construction projects over the past 11 years, summed it up like this: "Contractors should recognize that they are as much, if not more, in the service delivery business and not simply in product delivery." Service means providing attention and follow-up by the principals of the firm. "The project owner needs to know that it is valued by the principals of the construction firm, and actions speak louder than words." Often, "Principals sell the job and then disappear. When the shit hits the fan, they're nowhere to be found." Service also means "providing added value by working together as a team with a shared vision and doing more than what is contractually required."

Unlike the public sector, it isn't easy to obtain information on what individual companies or institutions are doing in the realm of construction unless you have existing relationships and contacts that are directly plugged in that can keep you posted. Marketing to the private sector requires constant research and an aggressive sales effort.

This means that you have to do a lot of legwork to get a chance to get in the door; you must develop a well-crafted and targeted marketing strategy for each type of private-sector client you intend to pursue, and identify and connect with the middlemen and ultimate decision makers.

If you're interested in working for regional retail chains, for example, I advise you to develop a target contact list of retailers; find out which contractors, architects, real estate brokers, and vendors work in the retail sector; read real estate and retail trade magazines to get a feel for who is doing what; send out introductory letters and brochures; attend conferences where retailers congregate; prepare a strong sales presentation with concise leave-behind materials; and set up meetings with contacts and prospects.

Your marketing message should be powerful and well placed so that it builds name recognition and familiarity, which, in turn, eventually builds confidence and credibility. Here's where the words you used to create your Market Positioning Statement come in handy to help you establish your logo and powerful phrases to describe your company's attributes.

Your marketing materials, from the company site sign, stationery, and business cards to brochures and your Web site, must be polished and professional and should present a clear image of your company. Your customers may be large corporations or institutions who expect nothing less.

The prospect may invite you to make a sales presentation before deciding to hire you. This is perhaps your only chance to make a good impression and you must seize the opportunity! To ensure that you make a winning presentation:

- Find out who you will be presenting to and what the major project issues are.

- Send the right person to represent your company—most often, that will be you.
- Include your project manager or site superintendent in the presentation and allow him or her to go over the details of the project and answer questions.
- Rehearse your presentation to remove the lumps and bumps.
- Distribute leave-behind materials that reinforce your strengths.
- Make a follow-up telephone call to offer additional ideas and solutions.

Your goal is to be visible to your target audience. If they hear a consistent message about you from different sources *concurrently,* interest in your company builds so that half the battle is won by the time you walk in the door for your first meeting. Try to imagine this scenario:

1. A prospect receives your brochure in the mail with an introductory cover letter attached to it. The brochure is well designed and easy to follow and she "gets" what you do immediately. Your cover letter refers to your Web site, which she proceeds to visit.
2. What she sees in your Web site echoes the brochure in its attractive design and user-friendliness and provides more in-depth information about specific projects, awards won, client testimonials, and articles written by or published about your company.
3. You follow up with a telephone call that ends up as a courteous voice-mail message.
4. Then, by chance a week or so later, the prospect is browsing through several trade magazines on her desk and comes across your advertisement in one magazine and a short announcement about a project you won in another.
5. A month later, your prospect runs into an architect she knows who tells her about a project he recently completed with you as contractor and says good things about you.
6. "Out of the blue," your prospect calls you to follow up on your mailing and voice mail and to invite you to introduce your company or to submit a bid on the next project.
7. Even if it's a small project for your company, you decide to bid on it. You know that you must start at the ground level to get in the door.

FIGURE 3.1 *Three Essential Ingredients for New Business*

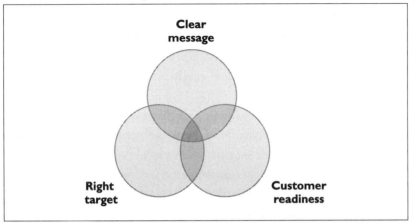

These events are not coincidental or serendipitous; they are a product of a smart marketing campaign. As one marketing-savvy contractor put it, "You make the cold call warm."

The more these three elements overlap—when you send a clear message to the right target who is ready to embark on a construction project—the better the chances of gaining a new customer. (See Figure 3.1)

Using the Right Marketing Tools

Many excellent books provide the small business owner in any industry with the basic tools of marketing. Here, I would like to focus on those tools that are most relevant and effective to the construction industry. What works for other industries (e.g., advertising during the Super Bowl broadcast or inserting discount coupons in magazines) does not necessarily apply here!

Your main goal is to send a clear message about your company's services and strengths that will grab the attention of your target audience and motivate them to call you or visit your Web site. To accomplish this, you'll need to choose your marketing tools carefully so that, blended together, you maximize the "buzz." These should include the following:

- *Promotional materials* such as brochures, letterhead, business cards, and postcards with a catchy logo or design that can be sent out as direct mail or distributed at a meeting, conference, or networking event

- *Professionally designed Web site* that is user-friendly, comprehensive, and up-to-date
- *Simple, well-designed advertisements* in trade magazines read by your target audience
- *Appearance in industry listings,* such as the *Blue Book,* which display your company profile and link the reader to your Web site
- *Press releases* to trade magazines and local newspapers about new and completed projects, awards received, new hires, promotions, and office relocations
- *Sponsorships* of trade shows, industry conferences, and seminars
- *Articles* written by you or team members on timely and interesting topics in your field and published in trade magazines and journals, with reprints you can send out to contacts
- *Presentations and exhibits* at industry conferences and seminars
- *Active membership in trade associations* in which you join committees and/or take a leadership or administrative role
- *Active participation as a board member* of a civic club, nonprofit organization, community board, political or social organization
- *Networking* at industry events and one-on-one meetings with industry colleagues and friends, vendors, suppliers, architects, consultants, construction managers, real estate brokers, attorneys, contractors, insurance and banking professionals
- *Associate membership in organizations that potential customers belong to* so you can develop contacts and become informed about their issues and concerns

Above all, deliver excellent performance, service, and follow-up and, once the work is completed, request constructive feedback. Your "debriefing" will create goodwill and show your customers that you are truly interested in meeting their needs. By doing so, you raise the odds that you will be invited to bid on the next job. If you win the second bid and perform successfully, you're on your way to establishing a track record.

EVALUATING BUSINESS OPPORTUNITIES: TO BID OR NOT TO BID?

Booking and executing profitable work hinges on the quality of the opportunities brought to the table.

If your marketing efforts are carefully targeted and consistent, you'll greatly improve your chances of attracting business opportunities that are of interest to you. Of course, building new business relationships

Biting Off More Than You Can Chew

ABC Construction Management is an old pro at the business. The company has been doing $2 to $10 million ground-up institutional and commercial projects without a hitch for almost 20 years. The past year, ABC accepted an invitation to bid on a project unlike anything it had ever done before—a $20 million rehabilitation/renovation of a low-income highrise apartment building that was fully occupied. Excited by the opportunity, ABC plunged into the project despite its daunting size, weak architectural drawings, and an incomplete understanding of the scope of work. It took less than five months before the company received a letter from the customer threatening to terminate the contract. What went wrong? ABC's project team had never worked on a project of this magnitude and scope. They hadn't had to deal with the tricky issues associated with coordinating dangerous work around tenants and communicating properly with the building owner. They made serious mistakes early on and undermined the customer's confidence. Fortunately, the company hired a consultant (us) to step in. We helped them to replace the entire project team, implement a project recovery plan, and get the project back on track. ABC eventually completed the project at a small profit.

Lessons Learned: If you want to properly evaluate a bid opportunity, you must be realistic about your firm's capabilities. Do you have relevant experience and expertise? Is the project in a market you know? Do you have the financial wherewithal/working capital availability? Do you have the right project management personnel? Does the type and size of the project fit in with the goals you outlined in your business plan? If you have doubts about the fit, you are better off taking a pass on the bid.

takes time and effort; project opportunities probably won't flow in your direction just when you need them. In the meantime, you'll be under pressure to drum up business and will be tempted to look at prospects or projects outside of your Target List.

Contractors get into trouble when they jump into projects to increase business volume without taking a cold hard look at what they are bidding on. They end up winning projects that are: too big (i.e., a $5 million-a-year contractor wins a $15 million job); in a new geographic area (e.g., a Florida contractor takes a job in Virginia); in a new line of business (e.g., a mason decides to become a general contractor); a cash-flow stretch. One prudent contractor quipped, "Volume is vanity and profit is sanity."

Before you go any further with bid preparation, you must take time out to scrutinize each bid opportunity *in detail* to determine whether the project makes sense given your experience, capabilities, financial resources, and the overall direction of your company. You also have to assess the risk and determine just how badly you *want* or *need* the job. In Chapter 7 under "Creating a Smart Bidding Strategy," we'll go into more detail about how to make that crucial bid decision.

More Reading on Marketing

Kubal, Michael, Kevin Miller, and Ronald Worth. *Building Profits in the Construction Industry*. McGraw-Hill, 1999.

Levinson, Jay Conrad, and Seth Godin. *Guerrilla Marketing Handbook*. Houghton Mifflin, 1994.

Society of Marketing Professional Services. *The Marketing Handbook for Design and Construction Professionals*. BNI News, 2000.

4

DEVELOP A DYNAMIC ORGANIZATION

So far in this book, we've concentrated on looking at market and industry conditions, assessing your personal and company capabilities and resources, establishing company goals and strategy, identifying your place in the market, and creating a plan to generate business opportunities.

Developing strategic business and marketing plans and creating a flow of business are great accomplishments, but they are only half the battle. Once you have your plans in place, you must develop a company organization with the capability to produce the results you expect, even if you are a "one-person show" at this time. Ultimately, the success of your business hangs on the competency and effectiveness of your organization!

As president, CEO, and founder, it's up to you to make sure your company is organized properly to take advantage of business opportunities, compete effectively, perform the work successfully, manage the risks, and protect your bottom line. Only you can turn your operation into a living, breathing organism that can deliver top-quality service to your customers.

SETTING UP YOUR "DREAM" COMPANY

You can determine the shape of your company at any given time by considering three important questions:

1. What is your business volume and anticipated growth?
2. What are your team's capabilities and experience?
3. What financial resources do you have available to invest in your organization?

Of course, developing a company that perfectly balances these variables is difficult. Jack Osborn, industry expert and attorney, observed, "It appears that contractors have a life cycle that allows some to form a small-high quality team that functions well . . . until expansion takes place, the quality erodes, success is diluted, and sometimes the company fails."

It's the old "chicken-and-egg" routine all business owners face—you need business growth to build cash flow and financial wherewithal, and you need cash flow and financial strength to generate business growth.

Your challenge is to *calibrate* company growth with your capabilities and resources. That means being cautious and conservative; build your organization slowly, brick by brick, as you work to bring in business, improve your financial position, open a line of credit, obtain liability insurance, develop bonding capacity, and build a team.

One of the best ways to ensure that business growth does not outpace your capabilities is to avoid taking business on simply because it comes your way. It's crucial that you evaluate each and every new business opportunity given your limitations before making the decision to take the plunge.

Any time you have a mismatch—when your company is too weak to handle increased business activity, or when your organization is too large to downsize in time for a market slowdown—you're headed for trouble. Arnold Marden, managing partner of Marden, Harrison & Kreuter CPAs, P.C., a construction accounting firm, advises emerging companies as follows: "You've got to take small bites or you'll choke. If you don't balance your revenue stream with the productivity of your team, you'll lose control."

Even if you're one of the fortunate people who are capitalized at start-up, it's not prudent to go all out and purchase expensive equipment or rent a big office and fill it up with staff, computers, and furniture just because you can, and not because you have the business volume to go with it. I've seen companies launch themselves with a lot of fanfare and go out of business a couple of years later. The construction market ebbs and flows, and high overhead costs will put you in the tanker when the market drops.

Conversely, when the construction market is strong, your organization may be inadequate to take full advantage of the opportunities that come your way. If you're unable to rise to the occasion because, for example, you don't have enough good people, you become vulnerable. Signs of trouble include making more mistakes than usual because you and your team are overloaded, which results in growing customer complaints. You'll scramble and stretch your financial resources to fill the gaps and perhaps get through the crisis—or perhaps not.

To create a workable organization, I recommend that you first design a "dream" company on paper that mirrors your overall goals and expectations for the business.

Even if you have no employees, create a simple organization chart with the basic job functions required to manage a contracting business. These would include business owner, salesperson, site supervisor, estimator, bookkeeper, and administrative assistant. At start-up, you may be wearing all these hats or sharing them with a partner or employee.

As your business grows, each of these functions will evolve to become someone else's job and eventually a company department: CEO, Marketing, Project Management, Estimating/Purchasing, Accounting and Financial Management, Office Administration, and Personnel. Figure 4.1 contains a sample organization chart for a company that is well along in its development.

Like your business plan, your organization chart at this stage of your business isn't "written in stone." It represents a framework that's fluid enough to embrace new ideas and adapt to changing conditions in and around the business.

Setting up a basic business structure will enable you to identify needed job functions and match existing personnel appropriately. The exercise will also help you identify gaps in experience, expertise, and support and start thinking about how you can minimize them. You can then prioritize your staffing needs and begin the recruiting process, one person at a time.

FIGURE 4.1 *Sample Company Organization Chart*

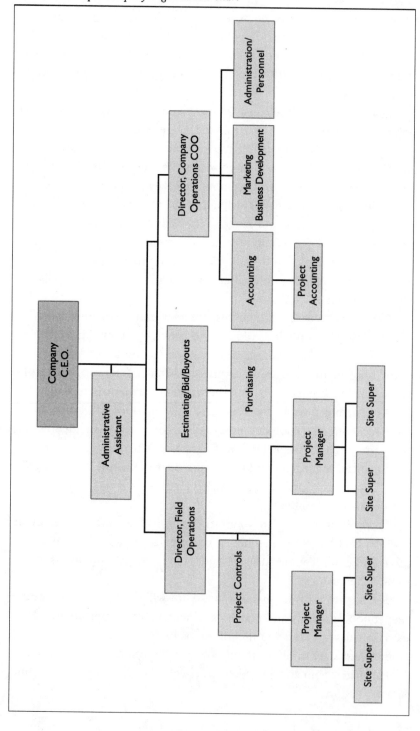

Remember that it's not a good idea to hire people for future needs *unless* you are hiring in anticipation of repeat work from a steady customer. Until you're ready, match basic functions with the capabilities of your current partners, staff, and yourself as best you can. Each person will wear several hats until you can afford to expand your circle.

Your organization is nothing more than the sum of its people, and it's the quality of your team that differentiates you from the pack. The trick is to hire the right people, place them in the right jobs, and provide

Growing Out of Business

Two project managers from a national construction firm got together one day back in the mid-1980s and decided to start their own general contracting company to work on commercial building projects. One of the principals was a real "rainmaker"; his business development talent led to a steady stream of new business opportunities and growing name recognition. By year ten, DEF had annual revenues of over $100 million. Fueled by their rapid ascent, the principals became even more aggressive; they branched out into virtually every type of building project in both public and private sectors, opened new offices, and began to add staff at a fast clip. By year 15, DEF had more than 300 people and annual revenues above $250 million. The principals believed that they had made it into the big leagues, competing with the largest construction companies for the most visible projects. Millions of dollars were poured into marketing, staff development, and frequent upgrades to office systems. DEF moved into new geographic territories across the country. By year 16, the cracks began to show: receivables had grown quickly and were staying on the books longer; cash flow was insufficient to cover expenses and payables were accumulating; borrowings hit the limit and bankers were unwilling to increase their support. By year 17, the company began to unravel; staff jumped ship and contracts were terminated. DEF went up in smoke, just like that.

Lessons Learned: DEF expanded way beyond its capacity. The principals were overconfident about their capabilities to sustain growth. Millions of dollars was frittered away on systems that didn't work and on new offices and staff in speculative geographic areas. Make sure that you are operating in markets you know, employ experienced and reliable people, and maintain sufficient cash flow and financial resources.

them with the knowledge, tools, and training they need so they can perform to the best of their abilities. We will cover these topics in detail in Chapter 5.

However you ultimately choose to organize your company, you must link and integrate the two sides of your business: field operations and office operations. In Chapter 6, we'll discuss in detail how you apply the glue by establishing effective company-wide standards, procedures, communication, and systems.

PUTTING YOURSELF AT THE TOP

If you originate from the design, engineering, or contracting world, where you've worked as an employee, chances are your expertise is in a technical discipline. If you've grown up inside your family business, you have a nuts-and-bolts understanding of various aspects of the operation. In either case, you probably have little or no experience leading or managing a team, planning and managing company finances, hiring and developing personnel, or handling marketing and business development.

If you're only one person, you are somehow managing to cover all these basic functions for now. If you've been in business for a while and have a partner and a few employees and associates, you're now managing a small organization of field workers and office staff. While you're busy chasing new business, your team is supporting you in the areas of administration, bookkeeping, estimating/purchasing, and construction operations. Everyone relies on you to make decisions; nothing gets done without your knowledge and approval; and you're aware of most daily activities and problems.

As you win more business and your staff grows, you realize that it's becoming increasingly difficult to be involved in day-to-day activities and decisions. You may feel that you're losing your grip on all the issues and potential problems. You're faced with a difficult choice: either you retain control and authority over everything or you begin to let go and empower your team to take responsibility.

Many contractor CEOs I know choose the former alternative; they insist on micromanaging all aspects of the business and, as a result, are unable to grow the company beyond a certain size. Burned out by the juggling, these CEOs can become dictatorial, temperamental, uncommunicative, or distrustful, which creates frustration and dissatisfaction among staff members. The result is a "revolving-door" atmosphere where people are constantly quitting and being replaced.

Tom Rogers, senior construction lender at Signature Bank who has been working with contractor CEOs for more than 25 years, commented, "When the business is doing well, the rising tide lifts all boats; when tough times roll around, all the cracks and leaks show. It's a constant cycle and that's why you [as lender] want to be involved with strong and experienced operators." It may take several crises before a CEO will stop in his or her tracks and reevaluate the situation.

The smart CEOs realize that they're primarily in the people business. They also realize that to maximize company effectiveness, they must allow employees to take ownership of their jobs. The team, their most important asset, walks out of the office every night, so it's critically important that the staff is motivated to return the next day, or there is no business to speak of.

A CEO of a fast-growing general contracting firm told me how she delegates and controls at the same time: Her job has three basic components:

1. Ongoing strategic planning
2. Customer relationship development
3. Cash-flow management

To make it feasible for her to concentrate on her job, she created a "steering committee" of decision makers that includes herself and the chief estimator, the chief financial officer, and the vice president of operations. They meet regularly to compare notes on what is going on in the market and to discuss particular company issues and strategies. Each member of the committee brings his or her perspective to the decision-making process and then communicates plans and enforces action in his or her department.

Another contractor CEO of a well-established specialty contracting business has developed a strong and loyal core team over the years, allowing him to evolve his role to that of "overseer, problem solver and negotiator."

If you discover that you're happier staying in the trenches in a technical capacity, you'll have to rely on a partner or hire a general manager, chief operating officer, or chief of staff to ensure that essential business functions are being performed properly. If this is not possible or desirable, consider going back to work for someone else before you get in too deep!

CREATING A POSITIVE ENVIRONMENT

As CEO, your goal should be to control the time you spend on the day-to-day details and focus on providing *leadership* and *strategic management* for your company. Leadership and strategic management mean communicating a positive vision and direction for the company and maximizing performance by inspiring a team effort toward common goals.

A CEO of a start-up company that is already doing well in its first year believes that attitude is everything: "As a leader, your glass must always be half full. Communicate your positive outlook and behave as if you are already there even if you're not."

The following actions will go a long way toward fostering enthusiasm, discipline, trust, respect, open communication, and collaboration among your team:

- Your staff can't read your mind. Communicate your plans and goals clearly and consistently.
- Convey optimism and resiliency even when you're having a bad day.
- Set priorities and explain how the decision-making process works.
- Provide your team with the tools they need to do their jobs well— set up standard company practices and performance measurements, and facilitate communication within and between office and field operations.
- Provide on-the-job training and send staff to seminars, conferences, and academic courses to improve or round out their skills.
- Create a forum for open dialogue, idea generation, and problem solving.
- Share your knowledge—be available as a mentor and as an advisor.
- Be sensitive to individuals' needs and issues.
- Review staff performance and provide constructive feedback.
- Don't let personnel problems fester.
- Give credit for good ideas.
- Provide incentives and deliver the rewards for performance.

- Establish guidelines for career progression.
- Organize off-site meetings and informal get-togethers where you have an opportunity to thank your team for their efforts.

DEFINING ROLES AND RESPONSIBILITIES

Here's another chicken-and-egg situation: you must exert leadership to develop a good team, and you need a good team to avoid getting caught up in the daily minutiae that prevent you from being a leader.

To gain and preserve your leadership position, you'll have to make every effort to maximize the effectiveness of your team. Besides hiring the best people in the business, set clear-cut roles and responsibilities for yourself and for each person in the company. Your role is to orchestrate their activities with your overall goals.

You can begin the exercise by writing brief job descriptions for each function that you've identified in your organizational chart, beginning with your job as CEO.

If you're a start-up business, it's likely that you don't have the in-house expertise and the number of people required to fill each and every job function, nor the business volume to support them. As you grow, you'll begin to hire or outsource positions and assign specific roles and responsibilities to individuals. To maximize the results of your efforts:

- Communicate the details of each job function so they are clearly understood.
- Review your expectations with each team member.
- Coordinate roles and responsibilities to minimize overlapping authority and duplicating efforts.
- Hold regular group meetings to maximize communication and collaboration among team members.
- Set agendas for group meetings, outlining issues to be addressed and assigning responsibility for follow-up.
- Expect staff to assume responsibility to carry out their jobs and be held accountable for the results.

COMPANY ROLES AND RESPONSIBILITIES

Date: _____

Summary Job Description:

President/CEO

Assigned To:

- *Provide direction and leadership*
- *Represent company to the world*
- *Forge customer relationships*
- *Develop team*

Project Management

- *Plan and manage the project*
- *Coordinate team activities*
- *Manage costs and schedule*
- *Communicate progress*

Field Management

- *Coordinate work*
- *Manage quality control*
- *Schedule deliveries, installations*
- *Manage safety*

Estimating/Procurement

- *Review new project plans*
- *Prepare cost estimates*
- *Finalize bids*
- *Purchase materials*

Financial Management/Accounting

- *Track company costs*
- *Manage cash/borrowings*
- *Prepare financial information*
- *Liaise with accountant, banker, surety*

Marketing/Business Development

- *Identify business opportunities*
- *Network with prospects, leads*
- *Implement marketing strategy*
- *Manage sales campaign*

Office Administration/Personnel

- *Manage company office*
- *Maintain files, systems*
- *Provide personnel support*
- *Purchase equipment/supplies*

SURROUNDING YOURSELF WITH RESOURCES AND SUPPORT

Staff Outsourcing

It can be very convenient and cost-effective to outsource certain job functions such as marketing, public relations, bookkeeping, and payroll/benefits administration.

Instead of hiring a full-time person to do your marketing, for instance, you can hire a part-time business development person who knows the ins and outs of the industry. You can also outsource bookkeeping to your accounting firm or to a part-time bookkeeper and engage the services of a payroll processing firm.

For a small contracting company, outsourcing can be a real lifesaver. You can instantly acquire expertise and expand your capabilities without having to increase payroll. And when you no longer need their services, you simply discontinue the arrangement. To avoid headaches, review your outsourcing options carefully and thoroughly.

Of all the functions you can outsource, payroll is at the top of the list for the small contractor. Construction company payrolls are cumbersome, time-consuming, and prone to error. They are frequent (often weekly), subject to change as project staffing fluctuates, and vulnerable to audit. If your wage rates and payroll tax payments aren't calculated correctly and submitted in a timely manner, you could get into trouble with the U.S. Department of Labor and the Internal Revenue Service. To handle your own payroll, you need a system that can generate detailed and accurate reports in a timely fashion.

If you're small and don't have administrative staff, you'll do yourself a great favor by entrusting the payroll function to a company that does nothing but payroll. For a listing of payroll processing companies, visit *www.payrollprocessing.us/*. Choose a vendor who is familiar with construction payroll requirements and who can take care of all the tax reporting as well. The processing cost is a small fraction of what you would have to pay someone on your staff to do the same work. Some payroll companies even provide workers' compensation insurance at a lower cost than you would pay on your own.

Going a step further, it's possible to engage a firm to handle all your personnel administration functions, from payroll, workers' compensation coverage, and medical and other employee benefits to 401(k)s, recruiting, and training. These Personnel Employment Organizations, or "PEOs," charge a fee for their services, which is largely offset by the benefits you receive from obtaining better and less expensive medical insurance benefits.

Primary Advisors

Your team isn't complete without outside advisors and experts who are available to provide guidance and support as you need it. Whether your company is a start-up or established, you must be able to rely on a good accountant, lawyer, and insurance professional and count them as part of your team. At times, you may also need the guidance of a business consultant. Some advice here:

- *Seek out professionals who specialize in the construction industry.* You must employ an accountant who knows the ins and outs of construction accounting and will help you establish credibility with your financial supporters. Equally important, you must engage a lawyer who is an expert on construction contracts and dispute avoidance. Likewise, your insurance broker must be knowledgeable about the construction industry and the risk management products best suited to your business. Similarly, a business consultant won't be much help if he or she knows nothing about the construction industry. These professionals will provide you with invaluable advice and links to financing and bonding sources.
- *Your selection process should include references and an initial interview meeting.* You want to ensure that you're engaging professionals who are going to provide you with the level of expertise, attention, and service you desire.
- *Develop an open and honest relationship with your advisors.* They won't be able to advise you properly if you're not straightforward and forthcoming with information.
- *It is worth paying them for their services.* Good advisors can save you a lot of trouble and money.

Surviving the "Perfect Storm"

GG, Inc., a small public-works contractor, decided to bid on a $15 million five-year federal project that required teaming up with an architectural firm to provide "design/build" services. GG chose a design firm it knew nothing about except that the firm had done work for the particular government agency and had an office nearby. Upon contract award, and anxious to get started, GG prepared a joint-venture agreement using a template out of a book and opened a joint bank account without consulting an attorney. The agreement called for a 50-50 split between the two companies with no clear-cut lines of authority and no delineation of roles and responsibilities. Together, they were to manage several projects dispersed among five states. Managing the southernmost projects closest to the project owner, the design firm insisted on handling the finances of the joint venture and dealing with the project owner on the joint venture's behalf. In the meantime, GG hired a new director of construction who was too overwhelmed with his new job to mind the store properly. Little did GG know that the design firm was in financial trouble because of bad investments in unrelated business ventures and that it had been dipping into the joint-venture account to ease its own cash-flow problems. To make matters worse, the project owner suddenly terminated the contract because of poor design work. GG was left hanging with unpaid receivables, missing cash, and $12 million in lost revenue.

Lessons Learned: Never "get in bed" with a firm you know next to nothing about—do your due diligence with regard to credentials, integrity, intentions, and financial stability. Don't try to cut corners or save money by excluding a competent attorney and accountant from the joint-venture setup and negotiation process. Make sure you're involved with every aspect of the work and fully plugged in with the project owner.

Advisory Boards

In addition, you can create an advisory board and meet periodically to discuss and resolve important issues facing your company. The board could include one or more primary advisors and one or two friends or colleagues who are in complementary or other lines of business.

Alternatively, you can join a CEO roundtable or advisory board organized by an independent entity, which includes CEOs from different industries. These boards provide excellent forums for open and honest discussions of your particular issues, problems, and dilemmas and are great ways to benefit from the expertise and advice of other CEOs. Above all, you'll have people to talk to about your business. As the saying goes, "It can be lonely at the top!"

LEVERAGING YOUR CAPABILITIES WITH JOINT VENTURES, ASSOCIATIONS, AND STRATEGIC ALLIANCES

How can you take advantage of exciting new business opportunities if you're a small company with limited capabilities and resources?

You can expand your horizons without stretching your organization to the breaking point by teaming up with established companies that possess the management capabilities, market presence, and financial strength that you don't currently have.

One way to do this is to enter into a *joint venture*—a legal partnership between two or more companies that is formed for the sole purpose of carrying out one or more specific projects. For example, you could join forces with a partner who has the bonding capacity to do the work while you bring to the table specific expertise or a strong customer relationship. Or you could partner with a design firm to provide a complete "design/build" package to the customer.

Several serious prerequisites to success must be considered, however, before you enter into any joint venture:

1. The joint-venture partners must get along.
2. All possible issues should be identified and resolved before a deal is made.
3. Each participant's scope of work, role, and responsibilities must be defined.
4. Participation, profit, and expense splits must be well negotiated and understood.
5. Avenues for resolving disagreements and dissolving the joint venture must be clearly delineated.

Before you go down this path, please take the time to get to know your potential partner well (do your due diligence) and seek the assistance of a competent attorney to help draft, review, and negotiate the joint venture agreement.

A less constricting alternative is to establish a relationship or "association" with a larger company in which you play a limited role on a project as subcontractor or subconsultant. Here you limit your risk while gaining valuable experience that would otherwise be out-of-bounds for you. If you deliver your end of the bargain and build a good relationship with your partner, chances are you'll be invited to participate on other projects as well. This could be great for building a track record and for strengthening your cash flow.

You can also shore up your capabilities by forging mutually beneficial "strategic alliances" with your vendors and suppliers through which they provide favorable terms and special products or services to you in return for your business.

More Reading on Leadership and Building an Organization

Bossidy, Larry, and Ram Charan. *Execution: The Discipline of Getting Things Done.* Crown Business, 2002.

Covey, Steven R. *The Eighth Habit: From Effectiveness to Greatness.* Free Press, 2004.

Gerber, Michael E. *The E-Myth Revisited: Why Most Small Business Don't Work and What to Do About It.* HarperCollins, 2001.

5

BUILD A STRONG TEAM

A construction company is only as good as the people who work there. Your success hinges on bringing in and retaining the most talented and dedicated people you can find and ensuring that they're well matched to their jobs, rewarded for their work, and given opportunities to improve their knowledge and skills. The big challenge in construction, as one contractor described it, is to "find exceptional people who know how to get things done."

If you don't make this a top priority, you'll continue to manage the details of your business. Therefore, never skimp on your efforts to attract good people to your organization. Start slowly, hiring one person at a time, only when you have the work and the resources to afford bringing someone on board.

CHOOSING PARTNERS AND KEY EMPLOYEES

In the start-up phase, you may be tempted to bring in a partner who has proven experience rather than training someone from the ground up. Ideally, this person will have the right credentials and the technical, marketing, or business management know-how you don't possess so that "two heads are better than one." One of you will be the "inside guy" (managing the work), while the other will be the "outside guy" (developing business). You assume that you'll achieve together what you cannot individually. You're willing to lose your autonomy for the benefit of sharing the pleasures and burdens of owning a business (including the financial responsibility). Best of all, you'll always have someone to talk to!

Building a good partnership is a big challenge for anyone. Whether you have a family or nonfamily business, some of the basic ingredients to a successful partnership are as follows:

- Shared philosophical vision, approach, and business strategy
- Mutual respect and trust
- Similar work ethic and commitment to the business
- Minimal overlap of talents and skills (i.e., one partner sets up and runs the business while the other executes the work)
- Clear division of roles and responsibilities
- Understanding each other's jobs
- Mutual agreement on major aspects of the business: partner compensation; profitability goals; strategies for capitalization, debt financing, and risk management; compliance with laws and regulations
- A well-defined decision-making process
- A methodology for resolving disagreements quickly and professionally
- An ability to enjoy each other's company because you are joined at the hip most of your waking hours

Your best bet is to put the details of your mutual understanding into a "partnership agreement," "shareholders' agreement," or "buy-sell agreement" with the help of your attorney. The ground rules must be understood from the start regarding such important matters as roles and responsibilities, percentages of ownership, distribution of profits, resolution of disagreements, and dissolution of the partnership. If one partner leaves, it should be mandatory that his or her stock be sold back to you.

I know of several partnerships that are working well. One example that comes to mind is a husband-and-wife team who founded their highly successful business more than 25 years ago. They're still happily married and truly have fun working together. The secret to their success as partners is threefold: Their roles and responsibilities are very distinct, with virtually no overlap; they respect and admire each other's talents and contributions; and they slowly brought in other partners and diluted their interaction in the office.

Another example is a partnership of two business associates who have been running a highly successful specialty contracting firm for six years. I met one of them for lunch recently, the day before he was to to go to Italy on a two-week bike tour of Tuscany. He told me excitedly that he'll go away with peace of mind because he can rely on his partner to keep the business humming in his absence. The secret to their success, he told me, is that they work well together and share the same work ethic and philosophy.

If a partnership isn't going to work, it'll become apparent very quickly. You may find out that your partner is not as dedicated or reliable as you originally thought, or not so participatory or communicative as you would like, or that your partner has some personality traits or quirks you find hard to live with. Or worse, it may become apparent that you and your partner have dissimilar goals and objectives and have trouble agreeing on business focus and direction. If you don't take action to confront the situation quickly, the conflict will become impossible to solve and the downward spiral will begin.

Without a legal agreement that spells everything out, it can take years to dissolve the relationship, often at the expense of the business. While you're fighting it out, the business may remain at a standstill as your staff watches from the sidelines, waiting for the first opportunity to jump ship. And, as in any marriage, ending the partnership can be emotionally and financially painful.

Matters are further complicated when your partner is a spouse, relative, or friend and you're not working well together. If healthy venues don't exist for communication and problem solving, emotions get in the way of sound business judgment and the decision-making process becomes dysfunctional. This unpleasant situation can lead to company paralysis and malaise, or worse. In the words of Abraham Lincoln, "A house divided against itself cannot stand."

From my personal experience and from watching contractor clients go through "hell and back" with their partnership arrangements, I recommend that you think very long and hard before taking a partner, especially if the partner is connected to you personally.

An excellent alternative is to hire a key employee, a professional with the wherewithal to make a valuable contribution to your organization as a key executive without permanent legal ties. You would utilize your preferred recruiting channels to identify a well-qualified individual with complementary experience and talents and the right disposition (e.g., energy, enthusiasm, commitment, determination).

You would also provide incentives by creating clear-cut performance goals and linking them to financial rewards such as commissions, bonuses, or profit sharing. You should spell out all the terms of employment in an "employment agreement" to avoid misunderstandings later. If it turns out that the employee isn't a good fit, you can terminate the arrangement and move on. Also, think long and hard before you hire someone you can't fire (e.g., your brother, friend, cousin).

Breaking Up Is Hard to Do

HJ Contracting was founded by two best friends who believed they were a perfect match. One was a gifted marketer who brought in the business and the other was a hands-on field manager who pushed himself to the limit to complete the most challenging projects successfully. They were such good friends that they thought they didn't need a legal agreement. Their honeymoon lasted for a few years while the business took off. Somewhere around their sixth year, the field manager partner began to falter. He showed signs of burnout: poor health, hints of alcohol and drug abuse, depression, lack of focus. His best friend was heartbroken to see him falling apart. He tried every which way to help his partner and family until he realized one day that the situation had reached the breaking point—he discovered that his best friend had been stealing from the company. The legal separation was extremely painful for both and took years to complete. During that time, company morale dropped to an all-time low and the business stagnated. Eventually, the surviving partner replaced his old friend and moved on with his life. To this day, I don't believe he has fully recovered emotionally from the ordeal.

Lessons Learned: It's almost impossible to be objective and behave in a professional manner when emotions are involved. With relatives and friends, business and personal issues can become jumbled up if you don't make a concerted effort to separate them. The result: dysfunctional business relationships and decision making.

HIRING AND RETAINING THE RIGHT PERSONNEL

When you decide it's time to hire one or more persons, it should be because you have immediate staffing needs. Most likely, you just won a new project and must find a project manager, superintendent, or foreman to begin the work as soon as possible. You scramble to make calls to people you know who may be suitable for the job. You also call your contacts and it turns out that a friend of an old business colleague is available. You meet briefly and, based on a quick review of work history and references, offer the person the job. He accepts and starts working. Weeks or months later, you realize that your new employee isn't performing well and that no one in your office likes to work with him. Now you have a problem: how to get rid of him as painlessly as possible and begin the hiring process again. Sound familiar?

In a time crunch, it's easy to ignore your vision for a "dream" organization and just hire someone to fill an immediate gap. For the sake of expediency, you forget the fact that it's in your best interests to hire people who will become part of a stable core team and grow with you. You end up hiring people who don't know what they're doing and don't fit in. As one contractor put it bluntly, "Hiring bodies backfires."

The best way to avoid making mistakes is to set up a *recruiting process* that is followed every time you have a staffing need, regardless of time constraints. Here are the essential steps.

Identify the Specific Staffing Need

Following are a couple of examples:

1. You're a general contractor working on office interior renovation projects in the $1 million to $5 million range. You just won a $2 million bid to do an interior fit-out for a high-end specialty retailer. This award presents an exciting opportunity to develop a retail customer base, an area you are interested in pursuing. You're now looking to hire an experienced project manager who has strong technical skills and the capacity to manage this and other jobs with minimal support. You

prepare a bullet-point *job description* that can be tailored for a job posting or advertisement. Your job posting could read something like this:

Construction Project Manager

Seeking project manager with minimum 8 to 10 years of experience and excellent track record in high-end interior fit-outs for retail and commercial projects in the metropolitan area; subcontractor coordination, estimating, purchasing, communications, and computer skills a must; knowledge of Expedition software a plus.

2. You're also looking to hire someone to help oversee your office operation while you're chasing new business and managing field activities. Here you're looking for someone with *managerial* talent who doesn't necessarily need specific construction skills. You can define the job requirements more broadly to attract a pool of qualified candidates within and beyond the construction industry:

Business Manager

Seeking business manager with minimum 10 years of experience and excellent track record managing day-to-day activities of a business operation, including: business development, bookkeeping/financial reporting, office communication, payroll administration, office systems purchasing and maintenance; accounting or business degrees preferred.

The notion of hiring a person from another industry with a business background should not be dismissed outright. What I like to call *cross-fertilization* can be a big plus; you'll have a larger pool of qualified candidates to draw from, and the new hire may bring a fresh approach and new ideas to your company that have proved successful in another industry.

Attract Well-Qualified Candidates

Whether you're looking for a key manager, an experienced project manager, a young foreman, an office manager, or a bookkeeper, the trick is to identify a person who is *well qualified* for the job. You want to avoid hiring an overqualified person who becomes bored and unproductive sooner or later or an underqualified person who becomes overwhelmed by his or her responsibilities.

Of course, hiring good people can be a difficult process of trial and error. If you make a mistake (and we all do!), it may cost you dearly in poor work performance and even lost business. And you'll spend more time and money to replace that person.

In my experience, the easiest and most cost-effective ways to create a pipeline of qualified candidates are as follows:

- *Tap your own people.* Your best bet is to provide training and career progression to encourage your own staff to take on more responsibilities and new roles. You can pick your best people to eventually take key management jobs.

- *Referrals* from employees, business friends and associates, and other industry contacts. They may know of potential candidates who could be a good fit for your company. If you put the word out and encourage your employees and contacts to keep their eyes open, there's a good chance that they will identify viable candidates.

- *Job postings on your Web site.* If your company is listed in the *Blue Book* or in another industry listing, eager candidates may contact you directly. My company receives several unsolicited résumés a month from people who take the initiative to locate our Web site, read about the company, and contact us. We review the résumés and keep them on file until the day we need to think about adding staff. In the past few years, we have hired five of these go-getters and have been happy with our decisions.

- *Local/regional colleges* for entry-level and part-time staff. You can place job postings and even arrange to interview candidates on campus. You'll have access to students who are looking for summer internships, part-time clerical/administrative jobs, and part-time or full-time technical jobs. For entry-level construction staff, you can access a pool of new graduates in engineering, architecture, or construction management who are eager to obtain hands-on experience in the real world of construction. These book-smart young people are quick learners and will make valuable contributions to your company before you know it.

- *Marketing materials* (brochures, Web site) that convey the message that your company is an exciting and rewarding place to work and has high performance standards. A candidate who reads your attractive and well-written materials will be positively inclined toward working for you even before the interview.

Other recruiting channels include the following:

- Online listings and classified advertisements in local and regional newspapers and trade periodicals (you will spend a lot of time weeding out résumés)
- Job listing services offered by trade associations
- Temporary agencies for short-term or part-time administrative jobs (a good way to test someone before hiring)
- Employment agencies that specialize in placing project managers, site superintendents, project accountants, and other site personnel
- Executive recruiters who, for a sizeable fee, identify and place qualified senior project managers, project executives, or company officers (i.e., controller, marketing director, chief financial officer, chief operating officer)

Choose the Right Person

One of the toughest and most important jobs you have as CEO is to select the right people to join your organization. Unless your company is large enough to employ a human resources manager, I recommend that you not delegate the job to anyone, especially when hiring key administrative and project management staff.

Before you interview a candidate, review his or her cover letter and résumé carefully and make note of questions or inconsistencies. Call references if they are provided and then contact the candidate. If your first impression about the person's credentials and communication skills is positive, you're ready to arrange for an interview. Major interviewing tips include:

- Break the ice by talking about something that both of you may be interested in (e.g., a hobby or interest gleaned from the résumé).
- Carefully prepare interview questions that relate directly to work experience, education, and the job the candidate will perform.
- Allow the candidate to do most of the talking and ask questions about your company and the position.
- Arrange for the interviewee to meet other employees to determine if he or she is compatible with the immediate supervisor and other team members.
- Perform thorough background and employment reference checks.

Beware: gauging someone's technical skills and experience during the interview is not enough. Your job is to be highly sensitized to important qualities such as attitude, motivation, work habits, aspirations, and potential for growth. One contractor I know looks for two main qualities: "passion and aptitude."

You also must look for someone who will fit in well with your current team and company culture. After all, your new hire is expected to perform a specific job *and* be a part of your team. Your "radar" should improve as you become more experienced with the hiring process.

Once you have the new hire on board, don't take the person's presence for granted. You'll have to fight tooth and nail to retain him or her because your competitors are just around the corner ready to pounce, particularly when the market heats up. You can brace yourself by offering career progression, specific incentives or rewards for good performance, and on-the-job and other training venues.

Evaluating Performance

If you're just starting to hire employees for your company, you probably don't have a formal performance appraisal system in place. You're working very closely with your staff and are observing their activities daily, and your feedback is direct and immediate.

As your organization grows, it is in your best interests to develop a system for evaluating your people. Open and consistent communication

There's More Than Meets the Eye

L & M Construction Company is a general contracting firm with approximately $50 million in annual revenues and a 15-year history of "organic" growth (translation: up/down/sideways) led by a very hands-on CEO and sole owner. A couple of years ago, he decided to take a step back from the day-to-day grind and hire people to perform key management functions. He started by looking for someone who would fill the shoes of Chief Financial Officer and Vice President of Administration. Based on a recommendation from a friend and great references, he hired an individual with very impressive credentials, a former CFO of a division of one of the largest national construction companies who was considered "brilliant" and a "financial whiz."

The references also warned that the man had no management or people skills, but the CEO didn't pay attention. Impressed with the candidate's ideas about putting the company "on the map," the CEO handed over major responsibilities and allowed the CFO to run with the ball. He found out a few months later that the man was a "walking textbook" but clueless about the inner workings of a small company. Never once, for instance, had he taken a close look at receivables or verified costs in the field, nor had he understood the details of the accounting system. In no time, he had lost sight of the fact that the company was overborrowing and losing money. He also didn't know how to work with the team to get things done and soon alienated the entire staff. In the meantime, the CEO took his first long vacation in years and came back to a company in shambles. Now, two years later, he's still working on reversing the damage.

Lessons Learned: It's always a challenge to find key people who can adapt well to your "culture." In this case, the CEO brought in someone who couldn't function in a small company environment. Look for people who have strong credentials and bring professionalism to your organization, have learned the ropes in a company of similar size, and can work with and motivate staff at all levels. Before making a decision to hire, meet the candidate several times in informal settings (e.g., on the golf course, over lunch or dinner) to learn all you can about his or her personality.

about expectations and performance goes a long way toward developing a motivated and dedicated team.

It starts with the job description you fashioned before you interviewed a staff member. Once the person is hired, sit down together and communicate your overall values, goals, objectives, and performance standards and review the list of responsibilities to ensure that your employee understands the job. If compensation includes monetary incentives, spell out what your specific targets are and create *accountability* for performance. For instance, a project manager would be expected to manage the project properly and meet profit and schedule goals; an office manager would be responsible for efficient document management and purchasing of office supplies; the financial manager would be expected to maximize cash-flow efficiency and investment returns.

Once expectations and rewards have been communicated and understood, sit down with your employee periodically. In a positive and open exchange, point out the individual's strengths and accomplishments and identify areas requiring improvement. Set clear goals and an action plan with a timeline. Encourage and listen carefully to feedback from the employee because you'll probably learn something about your business you weren't aware of. Agree to review progress in three or six months. If you determine that the person isn't making an effort to improve, begin the process of termination before too much damage is done.

Make sure to put it all in writing in a memorandum or a performance appraisal form. You can create your own form. I have included an example on the following page.

COMPENSATING YOUR TEAM FAIRLY

Before you can make a job offer, you must be aware of the mode and level of compensation that is appropriate for the person you're hiring. His or her compensation will depend on title/function, level of responsibility and decision-making authority, education, and level of experience. Your goal is to offer a competitive compensation package to ensure that you're in the market for the best people available.

To find out what the prevailing salaries and wages are, look at surveys put out by local trade associations, chambers of commerce, and the Small Business Administration. You can also look up government

PERFORMANCE REVIEW

Name: **Review Date:**

Date Hired: **Reviewed By:**

Comp. History:

Current Job Function: Assistant Project Manager

Performance Ratings (5 = Excellent / 1 = Poor):

Management Skills	*Technical Skills:*
❑ Understanding project objectives	❑ Site inspection/quality control
❑ Sensitivity to customer needs	❑ Coordination of work
❑ Understanding job responsibilities	❑ Estimating
❑ Planning/organizing skills	❑ Project progress reporting
❑ Leadership/initiative/resourcefulness	❑ Cost reporting
❑ Teamwork	❑ Change order management
❑ Time management	❑ Scheduling/review
❑ Written documentation	
❑ Familiarity with company procedures	

Areas of Strength:

Challenges/Areas of Improvement:

Employee Comments/Plan of Action:

Employee Signature: _____ **Date:** _____

labor statistics prepared by the National Labor Relations Board (NLRB) and the U.S. Department of Commerce. For skilled trade labor, you must abide by union wage rates or prevailing wage requirements set by the U.S. Department of Labor. For administrative functions, you can call your neighboring companies to find out what they are paying for similar jobs.

To motivate your new hire to perform to the best of his or her ability, provide a compensation package that, at a minimum, includes timely pay (at least every two weeks) and nominal salary increases. In addition, you can offer a performance-based incentive program that features sales commissions, "merit" raises, bonuses, or profit sharing.

One successful contractor I know pays "spot" bonuses throughout the year to team members who beat project completion and profit goals. This type of incentive is specific, results-oriented, and immediate.

You should also consider offering fringe benefits above and beyond those required by law (Social Security, unemployment insurance, and workers' compensation). These include medical benefits, life insurance, pension plan, flexible work schedules, sick leave, and paid vacations.

Like it or not, you'll have to offer benefits that are comparable or superior to those of your competition if you want to attract and keep the best employees. Of course, employee incentives and benefits can represent a considerable expense for your company and must be shopped around thoroughly and structured carefully.

PROVIDING PRACTICAL TRAINING

Even if yours is a small operation with a handful of people, you must make efforts to develop your staff. By providing them with practical tools and training, they'll improve their skills and keep pace with new methodologies and technologies. If you don't, your company will fall behind your competitors.

The spectrum of training venues is vast; go online and Google "construction training" and you'll discover a variety of training opportunities for construction industry personnel at all levels. These range from on-the-job training and other in-house training to independent workshops and seminars, apprenticeship programs, and college courses and degrees. Here are some recommendations for in-house training that can be conducted with minimal expense:

- *On-the-job training.*
  If you're hiring your first employees, you'll most likely be their mentor and teach them what you know. They will work side by side with you in the field and follow your lead. As you grow your organization and acquire senior and entry-level staff, you can pair them up on projects. If you put an experienced project manager and a project engineer together, the senior person gains support on routine tasks and paperwork and the junior person learns the ropes from the project manager. You can also establish a job rotation schedule for a junior person so he or she becomes exposed to various aspects of the construction process. In my company, we expose project engineers to the business side of the operation as well by assigning tasks related to marketing/business development, research, and strategic planning.
- *In-house training.*
  Your own staff can conduct training in specific areas of expertise. One successful contractor provides "free lunch" seminars weekly to all employees. The CEO and other key people talk about company culture, philosophy, communication, and other "big picture" topics. Project management and field personnel review site logistics or safety issues, construction means and methods, new materials and technologies, paperwork management, project team collaboration, and dispute avoidance. You can also bring in guest speakers to talk about important and timely topics such as contract administration, risk management, and project accounting systems. Your speakers could be your advisors (attorney, accountant, consultant) or industry colleagues and friends who have particular areas of expertise (e.g., site safety, green buildings).
- *Off-site training.*
  Off-site training can range from government agency–sponsored workshops offered at no cost to expensive seminars and training programs provided by industry associations and private vendors. For a small company, I recommend you consider the following training venues:

- *Government agency programs* provide training in the areas of safety, estimating, scheduling, contracts, project management, marketing, and business management. In addition, certification and licensing programs are offered in many states.
- *Trade associations* offer workshops and training to member companies on topics ranging from management to means and methods.
- *Software companies* provide hands-on training in the use of their products.
- *Local colleges and trade schools* offer evening extension courses that may be relevant and appropriate for you and other team members.

- *Reference and software library.*

  From the moment you start your business, create and build a library of essential up-to-date reference books that you and your staff can use daily to perform your jobs. These books are useful tools for estimating, purchasing, legal contracts, contract management, construction means and methods, safety management, and project management. Your library can also include essential software to facilitate project scheduling, cost estimating, project management administration, project accounting, financial management, and project team collaboration. And, of course, you should have copies of your local building code handy in the field and at the office.

More Reading on Management

Pfeffer, Jeffrey. *The Human Equation: Building Profits by Putting People First.* Harvard Business School Press, 1998.

Robbins, Stephen P. *The Truth About Managing People . . . And Nothing But the Truth.* Prentice Hall, 2003.

6

MAXIMIZE COMPANY EFFICIENCY

If you've started your business and are busy bidding on work and executing your first project, you already know that the construction business is information- and paper-intensive. The moment you bid on that first project, you're bombarded with information that must be analyzed and processed. The moment you sign a construction contract, documents begin to flow and don't stop until the project is completed.

Every step in the construction process requires analysis, planning, monitoring, reporting, record keeping and decision making. This produces a considerable amount of information and documentation that must be organized and communicated effectively within your company and to the rest of the project team.

You also know by now that construction is a people-intensive business. To execute a project successfully, your team must work seamlessly together and collaborate with all parties involved, including the end user, customer, prime contractor, subcontractors, architect and engineers, suppliers, vendors, and consultants.

In Chapter 4, we set up the basis for effective communication by creating your "dream" organization and assigning roles and responsibilities for each job function. In this chapter, we'll focus on setting up a business operation with an integrated system of practices, procedures, communication, and record keeping. In Chapter 7, we'll review effective strategies and procedures that you can use to control the progress and outcome of your projects. By setting up and integrating office and field operations, you help your team to work smarter individually and collectively.

The more efficient your organization is, the easier it will be to reach your ultimate goal: *to build a portfolio of successful construction projects that add to the bottom line.*

CREATING YOUR OWN "SYSTEM"

Even if you're only a two-person operation at this time, I urge you to begin to establish a *system* of managing information and communication for your company. You have too much at stake to "play it by ear," especially as your company grows and you expand your team.

If you don't begin to manage communication flow from the start, you will be depriving yourself of timely, complete, and accurate information with which to make important decisions. And your team will be forever scrambling to get the work done at the risk of making costly mistakes. Even worse, you will find yourself without the documentation required to substantiate extra costs or resolve misunderstandings and disputes.

Of course, each construction company is unique and has its own way of conducting business. When I refer to "system," I am talking about a mode of operation that works for your company specifically. The degree of rigidity that you build into your organization depends on the type of work you do and the kind of customers you have; the more projects you do simultaneously, for instance, the more controls you should have in place.

This also depends on your personality and inclinations. I know two successful trade contractors who have each been in business for about 20 years: one, a former engineer, is a stickler for organization—he "needs a structure and system to survive"; the other runs his company with few hard-and-fast rules—his core team members grew up with him in the business and they know exactly what he wants.

SETTING UP AN OFFICE

Let's start with your office, the command center for your business. Your setup should include the following:

- Telephones, fax, copier, and networked computers with file-sharing and e-mail capabilities
- A central filing system for all paperwork, from project documentation to financial and other records
- A basic billing, job costing, and financial record-keeping system
- A procedure to track receivables and payables
- A payroll procedure and employee time sheets and payroll and expense reimbursement forms
- Company stationery, business cards, and basic marketing materials
- Contact lists for customers, vendors, suppliers, and subcontractors
- Project contact lists
- A personnel contact list with emergency telephone numbers

At the very least, you should be able to contact people, share information, file documents and records, process payroll, bill for work performed, and keep track of money flows.

The next step is to develop basic forms or templates for routine correspondence and reporting so that your team members do not have to "re-invent the wheel" every time they send a letter or write a report. By standardizing your documentation, you simplify routine paperwork for everyone. Typical forms would include a bid decision worksheet, cost-estimating worksheet, supplier telephone quotation, subcontractor prequalification form, submittal, request for information, transmittal, daily activity report, labor report, change order request, progress report, meeting minutes, payment requisition, cost report, sample contracts, and standard notice letters and memoranda. Samples of many of these forms are in Appendix G. You can customize them to suit your needs and "brand" them with your logo just like your stationery and brochure.

Your business development effort can also be facilitated by creating prospect lists and standard marketing letters for each prospect type. You can create flyers for special offerings, other company promotional materials, and a simple advertisement that can be placed in trade magazines and industry listings. All these materials should be easily accessible to you and your staff as loose-leaf sheets or, better yet, as digital documents stored in shared computer files.

ESTABLISHING STANDARD PRACTICES AND PROCEDURES

The word *productivity* is typically associated with manufacturing or another process business in which people follow a specific sequence of work and standard practices and procedures to create or assemble a product. Productivity is the extent to which the operation achieves optimal quality and cost and time efficiency. The more productive the operation is, the higher the profit margin per unit. In construction, productivity is usually associated with the amount of work a worker completes during a period of time.

I would like to apply the concept of productivity to your operation as a whole. Despite the fact that every construction project is unique, there's a sequence of events that must occur in order to bid on and execute any project successfully. The *construction process* starts off with evaluating bid opportunities, making a bid decision, estimating and pricing the project, signing the contract, and establishing project quality, schedule, and budget targets and controls. This is followed by purchasing and storing materials, mobilizing field personnel and equipment, executing the work, monitoring project costs and cash flows, managing performance, and closing out the job. All these activities are multilayered and involve everyone on your team and several other parties (e.g., customers, subcontractors, architects, suppliers).

Assuming that you have construction experience, you are aware that there's only one right way to execute a project successfully and know instinctively what procedures must be followed. With a small team, you can communicate informally how you expect things to be done and control the results because you're in the trenches with your team every day.

Once you start working on multiple projects with an expanding core team and labor force, it's no longer possible for you to be directly involved with every detail of the construction process on every project. Your attention must shift to establishing and communicating your approach and methodology so that your team performs accordingly whether you're there or not.

This means developing your policies and procedures and putting them on paper in a user-friendly format. The sooner you begin to build this foundation of shared knowledge, the better. The benefits cannot be overemphasized. You will:

- Create important communication tools and guidelines for decision making and execution.
- Ensure that important functions are performed the same way every time.
- Help your team members perform their jobs with consistent accuracy.
- Reduce procedural mistakes and costly errors.
- Make it possible for your staff to work efficiently as individuals and as an integrated team.
- Facilitate employee training and help new hires assimilate to your way of doing business.
- Raise comfort levels and build credibility among your customers and bank financing, bonding, and insurance providers.
- Develop your company's culture and identity.

Right now, you're probably thinking, hold on a minute! I barely have my office up and running! Relax, I'm not suggesting that you establish and impose procedures overnight. This is an evolutionary process that starts with your leadership, is shaped by trial and error, and develops with the participation of your team.

When you're ready, you can begin to build the backbone of your operation by developing some basic documents:

1. An *employee handbook* should provide your team with a clear explanation of your personnel policies (vacation, sick leave, holidays, medical coverage, employment, performance evaluation, termination); job descriptions for main job functions; hiring practices, and written documentation about how you comply with various laws related to equal opportunity employment; procedures for dealing with employee grievances; and information regarding employee training requirements and opportunities. Check Appendix H for a sample outline of an employee manual.

2. A *project management manual* should be utilized by both field and office personnel as a definitive guide to managing projects and making decisions. It should include procedures for every stage and activity of a construction project and for handling project documentation and reporting. A sample table of contents can be found in Appendix I.

3. A *project accounting procedures manual* links project costs to your general accounting system and allows you to forecast and compare actual costs to the original budget. Your accountant can assist you in establishing these procedures.

4. A *quality-control program* outlines step-by-step procedures and controls for managing the quality of your work.

5. A *safety program* describes in detail all safety regulations and what steps your field personnel are required to take daily to ensure site safety compliance by all project participants.

Be aware that, just like your business plan and organization chart, the internal system you create is not etched in stone. It is a fluid framework that can be modified periodically to accommodate new and better ideas and technologies.

DEVELOPING A SYSTEM FOR REPORTING AND COMMUNICATION

As you develop your *tools* for communication, you must also establish the *channels* of communication that will integrate office and field into one functioning whole.

A good system for communication is the glue that will hold your organization together and keep the work flow moving at an optimal speed, thus enhancing productivity. It's also the *only* way your company will be able to generate timely and reliable information needed to monitor performance and make important decisions.

As we touched on earlier, applying the glue begins with a basic office setup so that personnel in the field and in the office can contact each other any time, communicate through e-mail, share information, and collaborate on assignments and projects.

It must also include a system of reporting that flows easily from field staff to office personnel and back, and from you down to the field superintendent and back up to you. Following is an example of how this works:

1. Your field personnel are responsible for providing detailed information regarding work completed, hours worked, manpower utilized, materials purchased and delivered, changes and issues that impacted work progress that day. The information is organized into daily reports and faxed or e-mailed to your office daily.

2. The project manager monitors all aspects of the project in the field and reports to you directly on the project's progress. The report should compare actual to planned progress, costs, and schedule; identify current and potential problems; and recommend strategies for corrective action. The project manager also submits to the bookkeeper/controller/CFO current information regarding work completed, materials purchased and delivered, and change orders. Your financial person prepares the payroll, payment requisitions, and job cost reports, and updates the financial information.

3. You review the information you receive from the project manager and your bookkeeper/controller/CFO and make decisions on pending issues. You pick up the telephone or schedule a meeting to communicate your directives.

One way to organize your flow of communication efficiently is to assign a "traffic cop," an office manager or field engineer, to make sure that project information is received daily and distributed and filed correctly. To avoid unnecessary paperwork or duplication, provide the office manager/office engineer with a checklist and schedule for submissions that must be tracked.

As CEO, it's essential that you maintain direct communication with your team via face-to-face meetings. These may include weekly project meetings with your customer, subcontractors, and other project participants; follow-up meetings with field personnel to review progress, issues, and corrective-action steps; meetings with your bookkeeper or financial manager to review cash flows, job costs, and financial position; business development and strategic planning meetings with your partners and key personnel; and periodic meetings with your outside advisors.

This may seem like a lot of "facetime," but meetings are still the best form of communication if they're well structured. To ensure that you and your team don't waste time in lengthy and unfocused sessions, schedule

What You Get Is Only As Good As the Input

OK Construction is a general contractor that has worked almost exclusively on public-sector projects of up to $10 million during its 15-year history. Recently, the company won two sizable projects, each in the $20 million range. In an effort to prepare the company for the increased demands that will be placed on the organization, the CEO and his team began to take a close look at current reporting channels from the field to the office. He confirmed that daily logs, subcontractor requisitions, change order requests, and other routine information are being assembled by the Assistant Project Manager and then passed on to the Project Manager for sign-off, who then sends it on to the Senior Project Manager, the Vice President of Operations and to the Controller for final review and sign-off.

In addition, the Project Manager prepares the Monthly Project Progress Report with detailed project cost information and sends it through the chain of sign-offs as well. The CEO has discovered that this elaborate system of "checks and balances" is not working. It takes days or even weeks before the paperwork makes it through the system and is processed; each senior staff member involved is flooded with paperwork on a daily basis; and the last reviewer (the Controller) still finds errors on a regular basis and ends up sending the paperwork back for corrections.

The CEO and his senior team have concluded that the problem originates from the source; mistakes are made by the APMs and PMs. Why? Several are new to the company and haven't had the benefit of consulting a company procedures manual (because it doesn't exist) or receiving in-house training (e.g., on how the company analyzes and reports on project costs).

Lessons Learned: Your system of communication and reporting should not inundate your staff unnecessarily. Without practices, procedures, and training in place, your staff will continue to make errors.

them in advance and set clear agendas and strict time limitations. As follow-up, assign a team member to generate concise minutes or reports that document items discussed, decisions made, and actions to be taken, with assigned responsibility and deadline for each action item.

MANAGING YOUR COMPANY RECORDS

In construction, a good record-keeping system is nothing less than a *requirement*. Like any other business, you must retain records to monitor your work, build historical information, meet legal requirements, and report to your customer, banker, insurance and bonding companies.

In addition, as a contractor, you must create a paper trail that will support your actions in the field day to day and provide the ammunition you will need if a dispute situation arises. Any construction project requires the participation of many players who, more often than not, don't see "eye-to-eye" on everything. If you don't have your ducks in order, you won't be able to defend yourself when a disagreement arises.

This issue reminds me of a contractor who stood to lose more than $350,000 on a $1.5 million project because he didn't provide detailed information to back up requests for approval for reimbursement of extra costs. He walked into my office one day, plopped down a cardboard box containing some loose, wrinkled, coffee-stained papers, and said, "Here are my project files. I need help to prepare a claim."

To ensure that your records are complete, documents should be dealt with as soon as they are received and filed properly so that no one has to waste time looking for them. Your *permanent files* will include financial statements, tax returns, insurance policies, and corporate legal documents. Your *project files* will include separate folders for such items as payment requisitions, bills, purchase orders, subcontractor supplier invoices, change orders, cost reports, meeting minutes, progress reports, correspondence, and construction contracts. A sample detailed project filing system is provided in Appendix J.

Make sure you protect your files by using fireproof filing cabinets. I also suggest you to keep hard copies of e-mailed documentation on file and that you send a copy of your financial records to your accountant or bookkeeper on a regular basis. For your computer files, bring in an information technology (IT) specialist to advise you on how to protect your hardware, software, and data.

USING TECHNOLOGY TO STREAMLINE YOUR OPERATION

As soon as you can afford it, invest in office systems and software that will further streamline your team's efforts to gather, share, and report information. A number of excellent off-the-shelf software packages are on the market that, if used correctly, will make a big difference in the quality and accessibility of information. You can start off with user-friendly individual software packages for bookkeeping, accounting, cost estimating, project scheduling, and business development management. Examples include Intuit QuickBooks and Timberline (bookkeeping, accounting, estimating); Microsoft Project and Primavera SureTrack (project scheduling); and Microsoft Outlook, Microsoft Access, ACT, and GoldMine (business development management).

You can also purchase an integrated management information system that meshes cost estimating, procurement, job costing, project and general accounting, payroll, document control, subcontract administration, and project management controls and reporting. These systems also provide all the templates and forms you will ever need. Popular packages include Intuit Master Builder, Primavera Contractor, and Primavera Expedition.

For very large projects, several packages include elaborate project collaboration features. In the 1990s, Web-based project collaboration software was the buzzword in construction and many vendors claimed to offer the best products. After the fallout from the dot-com disaster, a few surviving firms continue to improve and refine their tools. Their systems make it possible for many project participants to work together

Current Trends in Technology: A Tech Guru's Perspective

Technology for the construction industry has rapidly changed in the past two decades. Twenty years ago, two-way radios, fax machines, and computers with word processing were the hot technology setup for small and medium-size construction contractors. Today, technology is so vast it has to be addressed based on three categories: communication and mobile devices; desktop applications; enterprise systems applications.

Communication and mobile devices include cellular telephones, mobile e-mail/telephone type devices (Blackberry/Treo), tablet PCs (personal computers), and PDAs (Personal Digital Assistants). With this technology, contractors can communicate using wireless applications for mobile discussions, scheduling materials deliveries, punch-listing projects, and collecting work-in-place quantities. These devices have the capability to quickly collect, store, retrieve, and transfer information. Basic information such as names, addresses, and telephone numbers can be stored and quickly accessed and transferred. More sophisticated handheld devices can be used to retrieve Internet e-mail, provide directions based on Global Positioning Systems (GPSs) and weather conditions, and interpret human handwriting. Information can also be downloaded or synced into desktop and laptop computers using wired connections and wireless radio waves. Pricing for these devices continues to decline as the technology continues to mature. Devices such as Blackberry, Treo, SideKick, and Palm have operating systems, computer applications, and Internet access, and operate on wireless service subscriptions that require monthly service fees. The largest benefit of these devices for contractors is the ability to collaborate verbally, electronically, and without delay. Because the dynamics of construction require the constant movement of the staff, these devices provide economical, reliable, and accurate means to collaborate efficiently. Whether a contractor is a "one-man band" or a large general contractor, mobile computing is affordable, easy to use, and mainstreamed in the industry.

Desktop applications are computer programs that are installed on laptop computers or on the more conventional desktop computers. The current cost of PCs has declined over the past decade and the memory, storage, and processing speed has increased exponentially. Software applications most used by contractors include spreadsheets, word processing, scheduling, cost control, estimating, and Internet browsers. Peripherals such as printers, digitizers, plotters, scanners, and storage devices are available as devices to print documents, collect information from paper drawings, capture and copy paper information to an electronic format, and save vast amounts of information in data formats for easy and quick access and retrieval. Most construction organizations install Local Area Networks (LAN) to form a client/server configuration to share applications and data among the office desktop and laptop computers.

Current Trends in Technology:
A Tech Guru's Perspective (continued)

The networks can be hardwired or use wireless radio waves through routers. When the systems are networked across multiple offices, Wide Area Networks (WAN) are configured, using dedicated telephone lines and wireless communications. The most popular networking software includes Microsoft and Novell.

The third category of technology is *enterprise systems applications*. Enterprise systems are designed to include all the required applications that are needed to efficiently operate a construction company. Typical functions include financial reporting, estimating, scheduling, and the ability to store multiple projects and company entities all in one database. The largest systems include Oracle Projects, SAP, and J.D. Edwards. Medium-size contractors have a selection of offerings from Primavera Systems, Timberline, Intuit Master Builder, and Deltek. Benefits of enterprise systems include collecting information from every department and summarizing the data into Key Project Indicators (KPIs) that signal management that problems exist.

For small contractors interested in moving from the "paper and pencil" age to basic technology applications, system recommendations include cell phones, QuickBooks (bookkeeping), Primavera Contractor or SureTrak (scheduling), Microsoft Excel (estimating), Microsoft Word, and an Internet connection with browser. For contractors generating revenues in excess of $5 million, more comprehensive systems are recommended that include Primavera Expedition (project management), Primavera P3 or P3e/c (CPM Scheduling), Intuit Master Builder (accounting), Timberline, MC2, Hard Dollar, WinEst, or Quest (estimating), and a computer network with Internet access with browser.

Contractors today must acknowledge that technology for business is no longer an option. To be successful in the competitive and risky profession of contracting, technology must be an integral part of the organization. Similar to hiring a competent staff and continually investing in their skills and talents, technology requires the same type of investment to stay competitive and deliver successful projects.

By Jay R. Several, ETRAC Solutions, Inc. (www.etracsolutions.com)

seamlessly from remote locations. If you participate in very large, complex, and lengthy projects (e.g., road infrastructure), chances are you may be required to plug into such a system. These include Autodesk Constructware, Meridian Project Systems, Primavera Systems Prime Contract, Citadon, and E-Builder.[1]

Obviously, the more complex the system is, the more it costs to purchase and the trickier it is to implement successfully. So beware:

- Investigate and carefully choose the products that best suit your immediate and foreseeable needs. Technology is always changing—there's no need to invest in complex software that your company won't be ready for in the near to medium-term.
- Make sure your staff receives the training they need to fully utilize the systems you purchase. Alternatively, hire someone who has learned the ropes at a prior job and can train others to use the system. Some contractors are willing to spend money to acquire fancy software but neglect to provide the training that usually comes with it. The likely result is that the software is either used improperly or left in a drawer.
- Instill in your staff a spirit of teamwork and cooperation and a commitment to overall project success. No collaborative product is going to work if project participants don't want to work together!

More Reading on Company Practices and Procedures

Civitello, Jr., Andrew W. *Construction Operations Manual of Policies and Procedures.* McGraw-Hill, 2006.

2004–2005 Contractor Productivity Survey Results. FMI Corporation (*www.fmi.net*).

7

PLAN FOR
PROJECT SUCCESS

When you sign a contract to per-
form any kind of construction work, there's no turning back—you're expected to deliver what you promised. You must complete the job on time
and on budget and produce the quality the customer expects or you'll
be out of the game. And to make it worthwhile, you'll do everything in
your power to live up to your profitability goals and minimize the headaches along the way. The faster you're in and out, the better.

Of course, this is easier said than done. There's no such thing as a
perfect construction project, where the plans and specifications are
crystal clear and complete, site conditions are ideal, the end user knows
exactly what he or she wants, and the parties involved work together in
blissful harmony. There's also no such thing as a "normal" construction
project in which everything goes according to plan; the unexpected always looms around the corner.

Even before groundbreaking and throughout the course of the
work, a vast assortment of issues will pop up to challenge you, including design flaws and discrepancies; unforeseen site conditions; funding issues; an ambitious completion date; changes to the project scope;
materials and labor shortages; delayed materials deliveries; differences
in contract interpretation; disagreement over change orders; uncoop-

erative project participants; poor communication; bad weather; a difficult customer or end user; underperforming subcontractors; antagonistic neighbors. And the list goes on.

Too many contractors allow themselves to be vulnerable and even defenseless when the curveballs come at them. Unappreciative of the benefits of *planning,* they prefer instead to plunge in headfirst without a clear-cut approach and methodology to project execution. Confronted with a problem, they run the other way, hoping it will resolve itself somehow. The inevitable result is the seat-of-the-pants and crisis management I described in Chapter 1.

It doesn't have to be that way for you. The best way to overcome obstacles is to establish a *framework* for planning and executing your projects. This framework will be based on proven best industry practices and your particular project management strategy and approach.

This chapter will provide an overview of the essential ingredients to successful project management: *planning, management, controls,* and *collaboration.* These basic practices are proven by successful contractors to be effective methods of controlling the progress and outcome of the work and protecting their bottom lines. One well-organized contractor insists, "Whatever is controllable must be controlled to the nth degree."

Going back to Chapter 6, successful project management is supported by company practices, procedures, and communications. Please take time out to complete the Project Management Procedures Workbook located in Appendix K. I designed this comprehensive workbook to help you review every aspect of your project management practices and identify where you can improve. You can also download it at *www.ConstructBiz.com.*

As I indicated in the Introduction, I leave construction methodologies, means, and methods to you. Many excellent reference books and textbooks are available that cover these technical details.

CREATING A SMART BIDDING STRATEGY

There's probably nothing more critical to your business success than how you estimate and bid on projects. Your profitability hinges to an overwhelming degree on your ability to understand the details of the scope of work and anticipate costs accurately; errors can lead to project losses and even business failure.

And yet, bidding is not an exact science; it is a combination of hard numbers developed from historic costs and real price quotes, and subjective numbers based on "gut feelings" about the risks of a project. Getting it right requires that you have bidding expertise in-house, an in-depth knowledge of the market, and sound practices and procedures that your team follows closely. It also requires that you have a thorough understanding of all your obligations under your contract documents, which will be discussed further in Chapter 8.

Here are the key ingredients of good bidding:

- No matter how busy you may be, take time out to scrutinize each bid opportunity *in detail* to determine whether the project makes sense given your experience, capabilities, financial resources, and the overall direction of your company. You're far better off bidding selectively than responding to every opportunity that comes in the door.

- Before doing the takeoff, carefully review the contract documents (i.e, plans, specifications, construction contract) and determine whether there's a match with your capabilities. Is the project size within your range of experience? Are you familiar with or have you worked in that geographic area? Have you done that type of work before? Do you have the required bonding and insurance? Are the project budget and schedule reasonable? Can you live with the contract provisions? Can you meet the labor requirements? Are there especially risky aspects of the project that require intense scrutiny, planning, and supervision? By answering these essential questions up front, you can assess the risk and decide how to handle it—do you refrain from bidding or do you put a price on your perceived risk and see what happens? At the end of this section is a Bid Decision worksheet that you can use as a guide *every time you bid.*

- Prudent bidding requires strong in-house cost-estimating capabilities and skills in contract interpretation. I frequently receive calls from contractors looking for someone to bid for them because "We're too busy." Contractors who outsource this part of their business are playing with fire. They often get into trouble once they're awarded the job because they don't know the details of their scope of work, have no strategy for execution, and are held hostage by a budget they didn't create.

- Establish detailed estimating procedures so that you and your team don't miss an important step in conducting the due diligence required. These include reviewing the plans, specifications, and general conditions; making a list of questions for the design team; investigating site conditions; scoping out the work thoroughly; sending out bid information to subcontractors and vendors for pricing; preparing a detailed estimate of materials and equipment costs based on your quantity takeoffs, historic cost information, and fresh price quotations from vendors and trade contractors; forecasting your manpower needs and calculating supervisory and labor costs; and allocating costs for job site and general office overhead. If you skip a beat and neglect to include an item (e.g., concrete), you will be in trouble even before the project gets off the ground.

- Use a consistent cost-estimate spreadsheet format that is organized properly according to Construction Standards Institute (CSI) industry cost codes and facilitates calculations. You can either create your own spreadsheet on Excel or purchase estimating software that also provides productivity and price databases. In addition to materials, labor, and equipment, be sure that your spreadsheet includes line items for general conditions, overhead, profit, and contingency. Also, don't forget taxes, materials delivery charges, permits and other fees, interest expense, and indirect costs such as general maintenance of tools and equipment.

- Ideally, obtain price quotations from subcontractors and vendors that you've prequalified based on their capabilities and track records with you. Pass on all pertinent project details and any changes that are communicated to you during the bidding process so that you can feel comfortable that they're providing you with up-to-date, accurate, and reasonable pricing. Try to avoid taking the low bid from a subcontractor or vendor you never heard of!

- Don't wait until the last minute before the bid date to put your final price on your bid. Give yourself adequate time to assess various factors that will impact your performance on the project, including availability and costs of materials and labor, the quality of the design documents, the size of the supervisory team required, work-quality expectations, schedule constraints, conditions on the site, and the reasonableness of the contract provisions. Allow for a contingency amount for unforeseen or unknown elements.

- Your profit margin should be based on your company targets and on the level of risk being taken relevant to your other projects. You're not doing yourself any favors if you add on an insufficient profit percentage to win a job; you're squeezing project cash flow and putting your company's capitalization at risk.

- You can expand your cost estimate to become a bid tabulation form that will show the original budget and subsequent buyout amounts. Final pricing for each line item can then be rolled into the project budget.

- Create a complete bid file with the details of your bid so that you can develop historic information to assist you on future bids.

- Know contractually what you are bidding on. Besides scoping out the plans, specifications, and general conditions, carefully review all the terms of your construction contract. Take particular notice of bonding and insurance requirements, payment terms, schedule requirements, changes to the work, default provisions, documentation, and notice requirements. Create a *contract summary* for easy reference by your project team. We'll go over this topic in greater detail in Chapter 8.

- *Don't* bid on the job if you:
  - Don't completely understand the plans and specifications.
  - Haven't had time to analyze the opportunity and risks thoroughly.
  - Complete the Bid Decision exercise that follows and realize the project will overextend your staff and financial resources.

PLANNING THE PROJECT

Many contractors don't place much value on planning because they believe it is a "waste of time" and creates a "paper mill." Instead, they focus on getting the job and, once it comes in the door, mobilize without developing a strategy to execute the project. They ignore the plain fact

THE BID DECISION

Date: _____

Project:

Owner: BID DATE: _____

Project Description	Experience/ Capability	Match? (Y/N)	Comments
Type of Project (i.e., residential, school, retail, hospital, office)			
Type of Bid (i.e., lump sum, negotiated, time and materials, unit price)			
Project Size Dollar value			
Project Scope Your particular scope			
Customer Owner for prime, general contractor for sub			
Geographic Location Your familiarity?			
Architect Do you know the firm?			
Special Conditions What are the tricky issues and conditions?			
Schedule Can you work with it?			
Budget Can you work with it?			
Labor/Equipment Requirements Union, nonunion?			
Cash-flow Requirements For initial mobilization?			
Bonding Capacity Do you have it?			
Insurance Requirements Can you meet them?			
Type of Contract Do you have experience?			

that projects that start off well are more likely to end well. Once they begin, they're literally gambling on the outcome.

Planning is *critical* to the success of your construction project because it creates a framework that helps reduce uncertainty. The process entails creating a sequence of events for executing, monitoring, and controlling a project based on your analysis of project characteristics and risks. If planning is conducted properly using a realistic estimate and schedule, a project plan will:

- Help define project objectives given time and cost constraints
- Provide a sense of commitment to the project team as they work toward common goals
- Provide a workable framework that the customer can agree to
- Improve the productivity and efficiency of each individual and the team as a whole
- Make it possible to measure project performance relative to targets and take corrective action
- Develop a team mind-set focused on anticipating problems and finding solutions
- Greatly improve the chances of project success

Choosing the Right Project Management Team

The biggest indicator of project success is the quality of your project team. If you're looking at a bid opportunity and don't have qualified project managers and site supervision available, you're better off passing on the bid this time. Putting on weak personnel or hiring people you don't know for a new project can spell outright disaster.

During the cost-estimating phase, you'll determine the appropriate level of staffing for the job so that you can formulate a budget for project supervision. The staffing level will depend on the type and size of project, the location and duration, and the level of complexity of the scope of work. Field personnel can include one or more project managers, site superintendents, foremen, and field engineers. Administrative and technical support will be provided from the office in such areas as project accounting and reporting, reviewing and processing change orders and payment requisitions, and document management.

You clearly don't want to understaff or overstaff a job. Understaffing a job leads to poor performance because team members are stretched and end up spending time putting out fires. Overstaffing is not efficient, either; you may have too many people stumbling over each other at a high cost.

Getting it right requires a real understanding of a project's scope, characteristics, and risks; having good people to draw from; assembling a team that has the right "chemistry" who can work well together; and making sure that your key project personnel are committed to the project from start to finish.

The most important role is that of project manager. His or her involvement begins during the estimating, bidding, and planning phases and intensifies during the execution and closeout of the project. He or she should possess the right level of experience and technical competence and demonstrate effective coordination and decision-making skills. The project manager should also be planning-oriented, be able to communicate effectively, understand and appreciate cost and schedule controls, possess good "people skills," and be able to represent the company to the customer, architect, prime contractor, construction manager, and other project participants. Above all, the project manager must have the management skills to motivate and direct the project team effectively. As CEO, you can empower the project manager to make day-to-day decisions and provide leadership and support on the critical decisions.

Another important role is that of the site superintendent. You will need an experienced person who has a solid track record scheduling and coordinating the work, monitoring the quality of installation and taking corrective measures, motivating and managing the productivity of the labor force, and ensuring that the job site adheres to all safety and security regulations and requirements. The superintendent should also possess good "people skills" that make it easy for him or her to communicate with all project participants, from the project owner to laborers and materials deliverymen.

Selecting Qualified Subcontractors, Vendors, and Suppliers

In a scramble to get the job bought out, don't forget to take the necessary steps to prequalify your subcontractors before finalizing business arrangements. Ask each candidate you haven't yet worked with to com-

A Low Price Can Spell Trouble

One general contractor told me a story about how she didn't follow her own rules and is still hurting from the experience. She signed up an electrical subcontractor without conducting a background check or requesting a bond because his price was low and he was recommended by the project owner. Long story short: The electrical contractor failed to perform and was terminated. In retaliation, he went to the U.S. Department of Labor to complain that his employees had not been paid for work performed. The agency is still withholding almost $1 million from the general contractor until the matter is resolved.

Another general contractor fell into the same trap: he subcontracted a large portion of his work to a structural steel fabricator and erector because the company's price was a lot lower than the others. He visited the subcontractor's shop but never checked his references. The steel contractor did shoddy work that required frequent inspection and correction, which delayed the project and put the project owner on edge.

Lessons Learned: Choose your subcontractors from a list of firms you have prequalified; check recent references; and beware of a significantly lower price.

plete a prequalification form with information about company history and track record, safety record, financial and bonding capacity, and references from customers and banks. Get on the telephone and double-check the information provided. If you deviate from the process, you may find yourself in deep trouble on a project because a subcontractor fails to perform.

Creating a Project Plan

Immediately upon bid award notification, you and your project team must devote quality time to formulating a detailed project execution plan. Although your plan will be grounded on basic practices that have worked for you in the past, it should be tailored to the realities and unique demands of that particular project. Your plan should include several important components.

The Project Schedule

Many contractors are overwhelmed by the demands placed on them at project start-up and don't bother to prepare a project schedule unless the customer requires it. In that case, they tend to outsource the task and file the schedule in a drawer to gather dust.

The project schedule is a *key management tool* and you would be foolish not to utilize it. As the saying goes, "Time is money"; delays mean lost profits. The schedule enables easy visualization of progress and potential stumbling blocks. Whether it is in the form of a handwritten diagram, a bar chart, or a sophisticated critical-path-method (CPM) project schedule, you must map out the flow of work required to complete your project on time. The CPM schedule can be "cost-loaded," whereby the budgeted costs for the various activities are included. It can also be "resource-loaded" to show the utilization of manpower throughout the project.

If you're a general contractor, your schedule will incorporate the schedules of your subcontractors and suppliers. If you're a specialty contractor providing the same service on several projects, you can create a master schedule that tracks your work flow across projects and helps you to move resources efficiently from one project to another.

Your project team must review the scope of work thoroughly and create a baseline schedule that outlines a detailed sequence of events, important action items, delivery dates, and critical milestone dates set by the customer or other parties to the project. Then, to facilitate project kickoff, develop a 30-, 60-, or 90-day schedule that outlines the most immediate and important priorities.

Continue that train of thought throughout the project by requiring your project superintendent and subcontractors to prepare a simple (even handwritten) *two-week look-ahead schedule* every week so they remain focused on planning and coordinating each and every step of the work.

By taking this exercise seriously up front, you'll:

1. establish a game plan for the project team to follow;
2. uncover potential issues and find solutions early on;
3. communicate your plans to your customer, subcontractors, and other project participants;
4. set the stage for effective schedule control.

The Manpower Plan

As mentioned previously, the manpower plan can be incorporated into the project schedule to map out manpower utilization throughout the project. Manpower planning enables productivity factors to be determined based on the number of individuals who will be required to perform given tasks within the time parameters established by the project duration for those tasks. The plan will establish whether overtime will have to be incurred and at what cost.

The Project Budget

The project budget is another essential management tool that you can't do without; it's critical that you establish a system to track changes to your original bid so you can take appropriate action to preserve your project profit goal.

Upon contract award, set up a project budget that will be used throughout the life of the project as a reference point for gauging project costs. Using your detailed cost-coded estimate as the foundation for your budget, you'll add information about line-item buyouts ("committed costs") and change orders and compare the revised budget with your estimate to identify what line items are over budget and under budget. Combined with the project schedule, your project budget is also the basis for your project cash-flow forecast in which you determine how expenditures will be incurred.

The Safety Plan

Your safety record is of utmost importance to your company and it's up to you to make sure that all requirements are met. Therefore, it's critical that you review site conditions and prepare a plan that addresses safety issues present on that particular site and protects adjacent properties and the public. The plan must assign a competent and trained person to be the site safety coordinator and stress that everyone on the site is responsible for preventing accidents. Please see the section entitled "Putting Safety at the Top" in Chapter 8 for details on implementing a site safety program.

The Quality-Control Plan

Your contract documents will dictate the level of quality the project owner expects on a project. If you have to correct poor workmanship, you'll lose worker productivity, incur extra costs, and hinder your project schedule. If at the end of the day you don't deliver a finished product in accordance with the contract documents, the customer will either refuse to pay you in full or file a lawsuit against you. Either way, your bottom line could be severely impaired.

Your plan should set quality standards in accordance with contract requirements, assign responsibility for plan implementation (usually by the site superintendent), and require good quality records and inspection reports. It should also outline procedures for inspection and testing by your site personnel and outside consultants, disposition or correction of deficient work, and resolution of disputes regarding tolerances or noncompliance.

The Site Logistics Plan

Based on your review of the plans and the project site, establish a site logistics plan that will delineate site access, your site office location, staging areas, materials and equipment delivery and storage locations, particular equipment required (i.e., a hoist), site protection (i.e., fences, barricades, scaffolding), and security. Have your plan ready for implementation on the day you mobilize and to serve as a guide for your field personnel throughout the project.

The Materials, Tools, and Equipment Logistics Plan

Your schedule depends on prompt delivery of the right materials in the right quantity and in good condition; one mistake and your work sequence and overall schedule will be impaired. Establish a procedure and assign responsibility for setting up a materials log, monitoring shop drawing completion and fabrication of materials, tracking orders, inspecting, testing and accepting deliveries, and protecting the materials from damage and theft. Remember that if you accept items that are incorrect or defective, you own them.

In addition, determine what tools and equipment will be required and make sure they're available and in good functioning order at the time they're needed. You must keep up a year-round maintenance and repair program. Labor productivity drops significantly when workers arrive at the job and don't have the proper materials, tools, and equipment to work with.

MANAGING PERFORMANCE

A good project plan isn't worth much if it's not carried out correctly. Given that you're working in a constantly changing environment, issues beyond your control will arise that can throw you off course even if you do everything right.

Therefore, your plan won't be effective unless you establish controls and reporting mechanisms to monitor performance and warn about impending problems. Armed with accurate and timely information, you have an opportunity to nip problems in the bud by responding promptly and taking corrective action as necessary. In addition, continuous job monitoring will minimize disputes and legal expenses. The punch line: *you must control the project instead of allowing the project to control you.*

Effective *management* and *controls* hinge on the following:

- *Top-quality site supervision.*

  Your safety, quality-control, manpower, and materials logistics plans are only as good as the quality of your site staff. Your project superintendent and foremen must be competent, experienced, well trained, proactive, and well versed on company procedures. They must be capable of planning field activities to maximize labor productivity, follow up on day-to-day details, and take immediate and effective corrective measures. They must alert you promptly about potential problems so that you can step in and make a decision if necessary.

- *Continuous schedule control.*

  Your project schedule is your critical path to successful project execution. Whether initiated by your company or someone else on the project, one small change can affect its sequence and duration. Therefore, you must update your project schedule regularly (daily, weekly, monthly as appropriate) to incorporate progress and changes to the project, including missed critical dates and delays (e.g., in materials ordering/delivery). The schedule report should clearly summarize any loss of time and identify corrective measures and costs. If you can't mitigate the delay, you'll have the ammunition to request an extension of time from your customer.

- *Accurate cost reporting and controls.*

  Effective cost reporting starts with practices and procedures that are transmitted to the grassroots level and up:

  - The foreman/superintendent should know how to gather accurate and detailed information about the work performed on the site and prepare reports that quantify labor hours, manpower distribution, work items, materials purchase orders and tickets, and equipment utilization.

  - Your bookkeeper must receive accurate project cost information in a timely manner to update your cost-reporting system and prepare payment requisitions.

  - The project manager must know exactly what his or her responsibilities are with regard to monitoring each element of the project, verifying work completed, reviewing cost and change order information, and projecting future costs to completion. He or she should be expected to prepare a summary cost report (you can provide a user-friendly template to make his or her life easier) that compares actual costs for the period with original and revised budgets, calculates the percentage of completion, provides information on pending and approved change orders, and estimates the cost to complete.

  - The report should provide you with a snapshot picture of project cost status as well as the details about each work item so that you can identify where you are over budget or under budget and take appropriate measures.

- *Prudent change order management.*

  It's critical that you obtain approval from the customer or project owner on change orders before you proceed with the work. To manage this process effectively, provide a detailed breakdown of the components of the change order (i.e., labor, materials, equipment) and the pricing for each component accompanied by quotes from suppliers. If the change order request is initiated by you because of differing site conditions, clearly state your reasons for the claim for extra cost and/or extension of time and submit a detailed cost breakdown. To keep track of all change orders on a project, list all requests on a change order log and track them until approved or rejected. Be sure to include approved and pending change orders on the project cost report.

- *Efficient document control.*

  Your project will inevitably generate a considerable amount of paperwork that must be tracked and managed properly to avoid delays in decision making. Your best bet is to create "logs"; these are spreadsheet forms that help your team track the status of an item, the action or decision required, the person or party responsible, and the requested and actual response time. Typical logs track contract drawings, change orders, transmittals, requests for information, materials order and delivery status, shop drawing submittals. and correspondence. If you log in a request for information (RFI) you sent to the architect, for instance, it should trigger a telephone call to the architect if you don't receive the response on time. If you use an off-the-shelf project management system, it'll simplify the process by automatically updating your logs every time you send out or receive a document.

- *Effective progress reporting.*

  Unless you have only one project going on at a time and are running it yourself, you'll need to establish a reporting system that channels accurate and timely information about each project so you can make prompt decisions. This system should include informal communication (i.e., daily telephone calls and e-mails to and from your project team to alert you immediately to potential issues) and formal communication in the shape of a straightforward *project progress report*. Your project manager will prepare a weekly/monthly report that summarizes the project status; describes the work performed during the period; compares the actual cost and schedule status with planned objectives; outlines current and potential problems and issues; and identifies important action items, deadlines, and assignments of responsibility.

Of course, the quality of the output you rely on to make important decisions depends *entirely* on the quality of the input. You can't ever be fully comfortable with the information you receive if you feel you don't have a competent team that understands your expectations and possesses the know-how and tools required to do it right—thus, the critical importance of hiring and retaining the right people (Chapter 5) and establishing effective company practices and procedures (Chapter 6).

Once the project is completed, it's *essential* that you sit down with your project team to review every aspect of job performance, including costs, schedule, quality, safety, labor productivity, profitability, management effectiveness, and customer satisfaction. Your objective is to identify what went wrong and understand the reasons behind the failures.

Without carrying on a thorough analysis at the completion of each and every job, you and your team won't fully understand what factors led to better-than-expected or worse-than-expected results. You'll move on to the next job without having learned from prior experience and may repeat the same mistakes.

WORKING WITH PROJECT TEAM PLAYERS

The fourth leg of effective project management is *collaboration.* As a contractor on a project, you're never working in a vacuum. You're only one project player among several—architect, engineers, customer, end user, prime contractor, subcontractors, vendors, suppliers—and you impact each other's activities and performance. Taking a defensive stance and assuming at project kickoff that you are surrounded by "the enemy" won't serve you well.

To simplify your life and enable you to deliver what you promised under your contract, you should initiate the collaborative process right away. You can do so by simply extending your hand to the other players in the spirit of goodwill. The fact is that all of you will be working with the same goal in mind: to complete the project successfully and profitably. By working together, you'll be able to anticipate problems, identify solutions, and steer the project in the right course. Work especially hard to encourage a collaborative and nonadversarial atmosphere of give-and-take with your subcontractors and prime contractor so that you minimize disagreements and payment issues. Put plainly, it is good business to develop relationships that transcend just one project.

If you're a general contractor, treat your subcontractors as well as you would want to be treated. One contractor I know goes as far as calling his subs "business partners." And that they are, for to succeed as a general contractor, you must develop a core team of qualified trade contractors and suppliers who can be relied on to perform good work at a fair price. They will be eager to meet and exceed your expectations if they are treated fairly and paid promptly. On this subject, one wise contractor echoed what many others have told me: "We're only as good as our people, subcontractors, and vendors."

If you're a trade contractor, you know that the prime contractor is in the driver's seat. You have to "take the crumbs," deliver the goods, and build a track record to be invited to bid again. Over time, you can build solid relationships based on performance, consistency, reliability, and trustworthiness. When it comes to money, don't be afraid to negotiate or even stand your ground when you're right. In the spirit of give-and-take, however, one specialty contractor advises not to "nickel-and-dime your prime." Another one jokes, "You don't have to be a pig. Pigs get slaughtered."

Relationships Can Save You

Ten years ago, PQ, Inc., was in its third year of operation as a small public-works general contractor when it got into serious trouble and was headed for bankruptcy. It started when the two partners bid on three projects of $1 million each and soon realized that they had underbid them by half a million dollars each, putting them in the hole for $1.5 million! (Why? They hadn't bothered to do cost estimates.) To make matters worse, the partners had no cash reserves or bank credit lines. Strapped for cash, they stopped paying themselves, borrowed money from family and friends, and opened more than 30 credit cards. The telephone rang off the hook with demands for payment by subcontractors and suppliers. Hoping to avoid panic, the partners took their calls, explained the situation, and promised to pay them regularly in small amounts. For more than two years, the partners kept their subs and suppliers fully informed and paid them in installments as promised. Thanks to the partners' prudent approach, not one sub filed a mechanic's lien against the projects. Eventually, PQ won a couple of new projects and the cash began to flow again. Fast-forward to the present: PQ is now a $50 million company and still works with a core group of subcontractors and suppliers who held steady during PQ's crisis.

Lessons Learned: PQ succeeded in avoiding bankruptcy by maintaining a dialogue with its subcontractors and suppliers and by eventually delivering on its promises to pay. The experience helped to forge long-standing business relationships that exist today.

More Reading on Project Management

Ahuja, Hira N., S. P. Dozzi, and S. M. Abourizk. *Project Management: Techniques in Planning and Controlling Construction Projects.* John Wiley & Sons, 1994.

Clough, Richard H., Glen A. Sears, and S. Keoki Sears. *Construction Contracting.* John Wiley & Sons, 2005.

8

MANAGE YOUR LEGAL RISKS

For most business owners, legal and insurance matters are the least fun part of doing business. Many of us literally avoid spending time looking at our contracts and insurance policies because the minutiae are tedious, dry, and intimidating.

In some circles in the construction industry, the word *contract* is even considered a dirty word. I often hear seasoned construction executives talk nostalgically about the good old days when "deals were done on a handshake." Many contractors still believe today that, once the contract is signed, it should be set aside as a mere formality and not spoken of again with the customer.

The reality is that construction is one of the most dangerous businesses around; even if you do everything right, you still run the risk of losses and even business failure because of contractual disputes, accidents and injuries on the job, and damage to property. By not confronting the details of your agreements and policies and by tiptoeing around the customer on contractual issues, you risk shooting yourself in the foot, or worse.

In Chapter 7, we talked about how we can manage construction risks by implementing smart project management strategies and procedures. In this chapter, I'll provide an overview of ways to limit your legal risks by actively managing your contracts, regulating job site safety and security, and protecting yourself from personal liability.

MANAGING YOUR CONTRACTUAL RISK

The construction contract is the game plan for the entire construction process. It determines what the project is, how it will be built, how much it will cost, and when it must be completed. If used properly, your contract is a key management tool for steering you toward successful project completion. If ignored, it can be the source of aggravation, loss, and failure. Proper management of the contract requires a thorough understanding of your rights and obligations and a proactive management approach from the bidding phase to final acceptance of the work by the project owner.

In this section, I'll take a broad-brush approach and walk you through the major land mines that await you as you manage your projects. Please consult your attorney for legal interpretations and advice specific to your contracts and to the jurisdictions in which you work. Be aware of the following.

You're on the Hook at Time of Bid

There are four types of construction contracts: *lump sum,* in which you provide a fixed price for the total amount of your work; *unit price,* by which you provide a fixed price for each item or unit (including labor, materials, supervision, insurance, profit, and overhead); *time and materials,* where labor hours and material costs are charged as incurred; and *cost plus* (or "guaranteed maximum price"), in which you charge for costs plus a fee. The riskiest form is the lump-sum contract because you're committing to an all-inclusive price.

It may take many months for the architect to draw up the plans and specifications, but you'll have just a few days or weeks to review. While you scramble to price out documents that are often unclear or inconsistent, it's possible to make mathematical or judgment errors that cause you to "underbid the job." If you make a big clerical error in bidding on a public project and are declared the low bidder, you may be able to withdraw or modify your bid depending on your state statutes. If you make an error of interpretation, however, you'll have to honor your price.[1] In the private sector, there's likely to be room for negotiation that may allow you the opportunity to uncover and correct errors; however, once you sign your contract, you're likewise committed to your price.

What happens if you enter into a fixed-price contract and then experience substantial price increases in a building material (e.g., steel, concrete)? Is there a legal means to recover your extra costs? The answer depends on whether you're working for a public- or private-sector project owner. In the regulated public sector, your ability to obtain relief depends on how strict your particular state statutes are and how cumbersome the claims process is. In the private sector, there's no legal obstacle per se. According to Henry Goldberg, senior partner of Goldberg & Connolly, a construction litigation firm, "A private owner can agree to it [a price adjustment] if he believes that it is in his best interests to do so . . . A sense of urgency in private-sector construction is always an issue, with construction loans and debt service looming as a backdrop to any negotiations. If 'time is money,' it may simply pay for the owner/developer to adjust a contractor's price to keep a job running smoothly."

Contracts Shift the Risks to You

The author of the prime construction contract tends to be the project owner or owner's representative. The owner will either draft his or her own agreement or adopt a standard boilerplate agreement designed by such industry associations as the American Institute of Architects (AIA) and the Associated General Contractors (AGC) of America. The legal agreement incorporates by reference the technical specifications and architectural plans and includes standard terms and conditions, general conditions, special conditions, and addenda. Together, they constitute the contract documents.[2]

According to the Construction Industry Institute (CII), "The ideal contract—the one that will be most cost-effective—is one that assigns each risk to the party that is best equipped to manage and minimize that risk, recognizing the unique circumstances of the project."[3] An "ideal" contract minimizes uncertainties and, therefore, disputes by being specific about the allocation of risk, a "win-win" situation for all parties involved.

The reality is that, in most instances, contracts tend to favor the author's interests. The owner, for instance, shifts as much of the risk as possible to the general contractor who, in turn, tries to pass on the risk to the subcontractors who, in turn, attempt to place risk on vendors and suppliers. This is accomplished through "incorporation by reference," in which terms and conditions of the prime contract "flow down" to the contracts that are lower in the food chain.[4]

Examples of project owner risk-shifting include the "hold-harmless clause," in which the owner is indemnified by the contractor, and the "no damage for delay clause," by which the owner avoids responsibility for causing delays on the project. An example of the general contractor shifting risk to the subcontractor is the "pay when paid clause," where the general contractor's obligation to pay the subcontractor depends on receiving payment from the project owner; if the owner runs out of money, both the prime contractor and the subcontractor are out of luck.

There Isn't Much Room to Negotiate

In the public sector, the contract is handed to you as a "fait accompli," or done deal. When you first read such a contract, you may feel that you're already "guilty before proven innocent"; the burden is on you to perform the work no matter what the circumstances are. In the private

---

Your Contract Doesn't Always Tell the Whole Story

R & R, a plumbing contractor, signed a subcontract with AZ Construction to perform plumbing work on a new gymnasium and cafeteria building for a public school district. The subcontract contained a general flow-down clause that required R & R to assume all the obligations and responsibilities that AZ had in its contract. Midway through the project, R & R filed a claim against AZ for delay damages; AZ had changed the sequence of the construction schedule, thus extending R & R's time on the job by six months. Little did R & R know that the prime contract required claims for delay to be made in writing to the project owner and architect within 21 days of the occurrence of the event that caused the delay. R & R's claim was rejected because it had submitted the claim after 60 days and therefore failed to provide timely written notice. R & R had to absorb the extra costs and barely broke even.

Lessons Learned: If you're a subcontractor, always request a copy of the prime contract to make sure you have a complete picture of your obligations and requirements.

sector, there's room to negotiate the terms of your contract, the extent of which depends on your bargaining power. The better your track record and relationship with your customer, the better the chances of reaching a fair and reasonable agreement.

If you're a subcontractor, the general contractor will hand you a contract that incorporates all or most of the terms and conditions of the prime contract in addition to your specific scope of work and other obligations. If you work regularly with the contractor on private projects, you've probably already hashed out a favorable subcontract. In the event that you are unable to negotiate, as is the case in the public sector, or if you are working with a general contractor for the first time, you'll have to grin and bear it. Jack Osborn, principal of John E. Osborn, P.C., a construction law firm, advises that "understanding the subcontract, knowing how to navigate its limitations, and keeping in close and effective communication with the general contractor go a long way toward consummating a successful business relationship."

Contract Rules Are Not Made to Be Broken

TV Metals is a small, established specialty contractor that won a contract to fabricate ornamental railings on a large project for a government agency. Soon after TV began the work, the architect completely redesigned the railings, thus doubling the scope and cost of the work. Eager to please the general contractor, TV immediately began fabrication without waiting for written approval for additional material and labor costs. To make matters worse, the prime contractor didn't submit the change order request to the project owner on behalf of TV. As a result, TV got into a serious cash-flow crisis that threatened the business. Luckily for TV, the project owner (in this case a government agency) intervened to make sure that the general contractor submitted the paperwork.

Lessons Learned: Your contract says that you must obtain written approval before proceeding with changes or additions to your scope of work. The earlier you seek resolution on extra costs, the better off you are. Your change order request should include a detailed breakdown of labor and materials costs and any schedule impact, with backup price quotes. Don't skip this important step because you feel intimidated by your customer. If you proceed at your own expense, there's a probability that you may never be paid in full.

Of course, you don't have to accept unfair terms and conditions—simply do not sign the contract. This alternative is possible for contractors who have enough strength of mind and pocket to walk away from a bad deal.

The Burden Is on You to Substantiate Your Claims

The contract requires you to deliver your work in accordance with specific quality, schedule, and budget requirements. If you encounter changed site conditions, changes in design, delays, and other issues that preclude you from meeting those expectations, you must provide written notice and deliver documentation (your "claim") to support your request for additional compensation or for an extension of time. If the change is initiated by the project owner, you must prepare a change order proposal that becomes a change order when signed by the owner. In either case, written notice must be made within a specific time frame and your documentation must be detailed, accurate, and fully substantiated. You must also obtain written approval before proceeding with the work. Compliance must be taken *literally;* if you miss deadlines or neglect to provide acceptable backup information, your window of opportunity to obtain relief will be closed.

Your Contract Is All You've Got

Once you sign your contract, embrace it as the best management tool available to protect your company from crises, lost profits, and litigation. Embracing it means:

1. Reading the contract thoroughly before bidding
2. Knowing exactly what your rightsand obligations are
3. Finalizing contracts with subcontractors, vendors, and suppliers
4. Actively managing your contract during construction
5. Avoiding litigation

Read the contract documents.

The contract documents constitute all your responsibilities on a project so you should know exactly what they are before you make your bid decision.

During the bidding phase, review plans, specifications, and addenda thoroughly to determine the details of your scope of work and those of your subcontractors. Attend the prebid site meeting, conduct a prebid

Key Contract Terms to Watch Out For
("In Plain Language")

Time of Performance: Project start and substantial completion dates are set forth.

Owner's Right to Stop Work/Carry Out the Work: If you neglect to carry out the work properly, the project owner may issue a stop work order, step in to correct deficiencies, and back-charge you.*Indemnification:* You will be required to hold harmless all project participants against the consequences of any negligence or omissions committed by you and your subcontractors in the performance of the work.

Differing Site Conditions: If you encounter subsurface or concealed conditions not indicated in the contract documents, there are specific notice requirements and procedures to claim extra costs and an extension of time.

Changes/Extra Work: Make sure you are aware of the procedures for change order work. You will need to document the price and schedule impact of the change and submit for owner approval before you begin the work.

Claims: Claims for additional cost and time require timely written notice and proper substantiation from you. In most instances, you will be required to continue with the work pending resolution of the claim. For subcontractors, your recovery will depend on what the contractor recovers from the project owner.

Delays: You can claim extra compensation and time for excusable delays caused by circumstances beyond your control. A "no damage for delay clause" means that you will not be compensated monetarily for "customary" delays (i.e., normal weather conditions, coordination of subcontractors).

Liquidated Damages: Some contracts charge a penalty for each day worked beyond the substantial completion date.

Key Contract Terms to Watch Out For
("In Plain Language") continued

Pay When Paid Clause: A contractor must pay a subcontractor within a reasonable time of receiving payment from the owner. As a subcontractor, you can take action (i.e., file a mechanic's lien) if the payment is unreasonably overdue.

Default, Termination, and Dispute Resolution: The contract will outline the conditions for default and termination and what avenues are available for dispute resolution.

site investigation visit, and request written clarification of the architect for any questions you may have. Make reasonable assumptions about unresolved ambiguities. Remember that special documents take precedence over general documents and technical specifications overrule plans.

As a subcontractor, be sure to obtain a complete set of construction documents for your scope of work and request clarification from the prime contractor on all ambiguous items. Whatever you do, don't price the project until you're comfortable that substantially all your questions have been addressed.

In addition, implied in the documents is the expectation that you will interpret the contract in a reasonable and prudent manner based on your experience and expertise, your knowledge of normal trade practices (i.e., industry-accepted tolerances and variances), and adherence to regulatory requirements (i.e., local building code, Occupational Safety and Health Administration (OSHA) safety standards).

Know the rights and obligations of the contract parties.

Shortly after you're declared the winning bidder, the customer may wish to finalize the contract quickly and get on with the project. Caution! This is a crucial moment and not the time to provide a knee-jerk response.

With the guidance of your attorney, thoroughly review all terms and conditions, including the scope of work, payment terms, start and completion dates, notice requirements, insurance and bonding requirements, and clauses for changes, delays, differing site conditions, indemnity, penalties and incentives, claims, dispute resolution, default, and termination. If you're a subcontractor, request a copy of the prime contract to confirm that your scope of the work is defined as accurately and completely as possible.

In addition, become fully aware of the obligations of the other parties to your contract. The *project owner's obligations* include providing you with constructable plans and specifications; taking care of hazardous materials; providing site access; obtaining building permits and zoning compliance; providing owner-furnished furnishings, fixtures, and equipment (FF&E); making timely decisions; refraining from interfering in your work; making prompt progress payments; and funding approved change orders.

Before you sign your contract, you should determine that the project owner has sufficient funding in place and confirm whether there are hazardous substances or unsafe conditions on site and bring them to the attention of the architect and the owner in writing.

The *general contractor's* obligations to you as a subcontractor include coordinating the work; informing you of all project developments and schedule changes that can impact your work; submitting your payment requisitions timely; representing you fairly to the project owner; obtaining change order approvals in a timely manner; and making prompt payment for work performed and materials delivered.

Before you sign your subcontract, you should prepare a precise description of the work you will perform and the materials you will use and include it as an appendix to the contract. If you're not being paid timely, as a last resort, you can file a *mechanic's lien* against the property or public construction funds. Beware, however, that filing a lien can exacerbate your problem because it will halt payment flows from the owner to the prime contractor. The best prescription for prompt payment is to carefully choose the contractors for whom you work.

Finalize subcontracts and purchase orders.

Make sure that you prepare a contract for each trade and have each subcontractor sign it and provide proof of insurance and bonding (if you determine the need for it) before beginning work. If you rely on a handshake, you won't have control over their performance and risk liability for employment taxes, workers' compensation claims, and other contract liabilities. Your purchase orders and supply contracts for materials, tools, and equipment must be clearly defined with respect to product definitions and quantities, pricing, payment terms, form and date of delivery, and warranties.

Actively manage your contract to completion.

In Chapter 7, we reviewed the major aspects of effective project management and how proactive planning, monitoring, and controls help prevent legal disputes. In a world where contract changes are the norm, it's also crucial that you have a systematic approach to documenting *in writing* every aspect of your activities, and meeting all notice requirements. You run the risk of forfeiting your contractual rights if the administrative requirements of your contract aren't met in a proper manner.

Once you have been awarded the job, and before you start the work, it's worthwhile to take time out to summarize the key provisions of your contract. Your *contract summary* and a full set of contract documents should be located at the project site and at the office for easy reference. This simple step will facilitate your project team's efforts to administer the contract once the project is under way.

Effective contract management entails prompt and straightforward *communication,* oral and written, with the project participants, abiding by all contract notice provisions and procedures, and providing timely, accurate, and complete *documentation* to support your activities.

Your project manager must inform members of the project team of any current and potential issue before it becomes a bone of contention. You minimize your chances of entering into a dispute situation if you meet problems head-on, identify solutions, communicate with the other parties, and take prompt corrective action. According to Michael Zetlin, senior partner of the construction law firm of Zetlin & De Chiara LLP, "Dispute avoidance should always be on the front burner for contractors. Ownership must ingrain in project managers and superintendents the need to be on the lookout constantly for problem situations. Painful, prolonged, and costly disputes are avoided by recognizing problems as they arise and resolving them immediately."

Avoid litigation.

Most contracts dictate the procedures that must be followed to settle a dispute, including mediation, arbitration, and litigation, or a combination thereof. If the contract prescribes *mediation*—a mutually agreed-on nonbinding negotiation—you have an opportunity to resolve the matter quickly and confidentially among yourselves with the help of a mediator with whom both sides feel comfortable. If it prescribes *arbitration*—a mutually agreed-on binding negotiation—the process may be more cumbersome, but your arbitrator will be a seasoned professional with superior knowledge of construction who is approved by the American Arbitration Association (AAA). Many contracts require mediation as a condition precedent to arbitration. To learn more about alternative dispute resolution methods, you can visit the AAA's Web site at *www.adr.org* and print out a copy of *The Construction Industry's Guide to Dispute Avoidance and Resolution* (as of July 2004).

If the contract prescribes *litigation,* you have no choice but to go before a jury to defend your position. Litigation is the least desirable way to settle a dispute because it's extraordinarily time-consuming, expensive, and disruptive to your business and will not necessarily yield the results you desire. One attorney shared with me that juries typically don't know enough about construction to evaluate the merits of a construction case fairly and may even be biased by their personal experiences with home-improvement projects. At the end, you won't be fully compensated for your out-of-pocket expenses and won't recover the costs of your time and personal distress.

Please consult your attorney before initiating a dispute or responding to one initiated by another party!

MINIMIZING THE RISK OF LOSS

You may not consider insurance an exciting aspect of your business but, whether you like it or not, it's absolutely essential. Insurance not only provides protection from loss but it is required under most construction contracts. You have to grin and bear the cost of high premiums (while, of course, trying to reduce them by maintaining a sterling record) because construction is inherently a dangerous business.

The job site, in particular, is teeming with physical risks that must be dealt with head-on every day to avoid unexpected loss or injury. Risks include damage to or theft of materials, equipment, and vehicles; damage to the property itself; and bodily injury to personnel. One serious accident can severely damage your financial stability and even put you out of business.

As a contractor, therefore, you must establish a *risk management strategy*[5] that reduces your exposure to loss as much as possible. Your strategy should include a combination of risk avoidance, loss control, and risk transfer activities:

- You *avoid* risk altogether by deciding not to do certain types of work that you consider dangerous.
- You *control* risk as much as possible by taking action to reduce the incidence and severity of loss. As CEO, it's your job to set up and implement strict rules and training with respect to personnel safety, materials storage and security, and proper use of tools, equipment, and vehicles. Your personal commitment to a *zero-tolerance* site must be clearly understood and enforced by your project team.
- You *transfer* the risk that can't be avoided or controlled to an insurance company. You'll require *liability policies* such as commercial general liability, auto insurance, professional liability, and workers' compensation to insure you against your liability to third parties (i.e., employees, customers, general public). You also may need *property policies* such as builder's risk that cover repair and replacement of damaged or lost property and materials.

The least painful way to obtain insurance is to look for a good insurance agent or broker who can assemble a comprehensive package that meets your company's specific needs. Like your bonding agent and banker, your insurance agent should become a member of your inner circle of advisors, providing you with the latest news about the insurance marketplace and pointing you in the direction of new/better/less expensive insurance products. You, in turn, should make it a point to communicate any changes in your business operations that would affect your insurance needs. Open lines of communication lead to trust and confidence, which can only facilitate your efforts to obtain the best products and rates.

Another avenue is to enroll in special plans offered by trade associations. It's worth checking whether they offer more cost-effective plans than what can be obtained through a broker or agent.

PUTTING SAFETY AT THE TOP

For a construction company, safety is everything. You live and die by your safety record. Your company's "experience modification rating," or "EMR," is based on the number of accidents and incidents that have occurred on your job sites. A contractor who controls losses under a good safety program will receive a "credit modifier" of less than one. A contractor who has a record of incidents and accidents will have a "debit modifier" of more than one. Your safety record is so important that potential customers are using your EMR rating as a criterion for prequalification. In addition, your EMR affects insurance premiums: the higher the rating above one, the higher your premiums, especially for workers' compensation.

Safety starts at the top (with you) and should be an integral part of your company culture. You must initiate the effort to create a company safety program, delegate day-to-day responsibility for implementation at each job site, and monitor compliance very closely. Following are some steps you can take to facilitate your effort:

1. Design a safety program in accordance with Occupational Safety and Health Administration (OSHA) requirements. You can seek assistance from a safety consultant, an OSHA "Small Business Consultant" (at no charge), or a union safety specialist. Each job site should have its own specific safety program.

2. Before you begin the work, notify the project owner and the architect in writing about any job site safety issues you have identified and wait to receive instructions.

3. Assign a competent person (i.e., a foreman) who is trained to recognize hazards on the job site and who is empowered to correct unsafe conditions or stop work until unsafe conditions are eliminated.

4. Display prominently your emergency contact information at the job site.

5. Provide site-specific safety orientation sessions and require OSHA training and ongoing job site training, including "toolbox talks."

6. Enforce the program by removing and dismissing workers from the job site and by charging fines for noncompliance (i.e., for not wearing hard hats, safety glasses, not using equipment properly).

7. Offer incentives to reward workers who are making exceptional efforts.

8. Report accidents and incidents immediately as they occur.

9. Require in your subcontracts that your subcontractors notify you directly whenever they learn of an accident that resulted in injury or property damage.

10. Above all, p*lan and coordinate your work with safety as the top priority.*

PROTECTING YOUR PERSONAL ASSETS

We've just reviewed how to avoid disputes on your projects and how to minimize the impact of losses and injuries on your business. I'd also like to touch briefly on the obvious: you must limit your *personal liability* in the event of lawsuits, claims, and property damage. The potential for losses and litigation in the construction business make it essential that you shield yourself personally.

Many small contractors are sole proprietorships in which the owners and their companies are considered one and the same for tax and liability purposes. This is the simplest form of ownership and profits from the business flow directly to the owner's personal tax return. With this arrangement, the contractor assumes complete responsibility for any of the company's liabilities and his or her personal assets are at risk. This situation might be acceptable for someone who is working out of his or her home on small home-repair projects. It is not appropriate for a growing business with employees, offices, equipment, and construction contracts, where the risk of loss could be substantial.

Partnerships are similar to sole proprietorships because they are easy to establish and allow profits to flow directly to the partners' tax returns. They also don't limit personal liability; the partners are each responsible for the actions of the other partners.

The only way to protect your personal assets from legal action is to make sure your business is a corporation. A corporation is a unique entity separate from the person who owns it. It can acquire, hold, and sell property, enter into contractual agreements, raise capital, and be taxed and sued. It also has an unlimited life, beyond those of its shareholders. Shareholders are only accountable for their investment in the company and can be held personally liable only if they fail to withhold and pay employment taxes. The shield also goes the other way; the company is insulated from judgments against or liabilities of its owners.

There are various forms of incorporation, including the regular C corporation, the Sub-Chapter S corporation, and the limited liability company (LLC). Construction companies typically favor the Sub-Chapter S corporation (S-Corp), which provides the protection of a corporation but allows you to pass through company earnings to your personal tax return. You benefit from having one level of taxation and paying taxes based on your personal tax rate, which is lower than the corporate tax rate paid by C corporations. Please consult your attorney and accountant to establish a corporation or to review your current legal and tax structures to ensure that you are set up properly.

Of course, a corporation will not protect you from personal financial liability if you provide personal guarantees for bank loans and bonding. You're on the hook unless your company has the financial strength to stand on its own without personal financial guarantees from you and your partners.

More Reading on Legal Issues

Hinzie, Jimmie. *Construction Contracts.* McGraw-Hill, 2001.

Kelleher, Jr., Thomas J. *Common Sense Construction Law: A Practical Guide for the Construction Professional,* 3rd Ed. John Wiley & Sons.

Sabo, Werner. *Legal Guide to AIA Documents,* 4th Ed. Construction Law Library, Aspen Publishers, 1998.

9

ACHIEVE FINANCIAL STABILITY

When it comes to financial matters, many contractor CEOs are busy creating revenues and executing the work, and they let the bottom line fall where it may. They don't focus enough on trying to control costs and are hard-pressed to calculate their company overhead rates or forecast cash flows. It may take a crisis before a contractor "gets some religion" regarding the importance of financial planning and management. And yet, their success hinges on creating a profitable operation and long-term financial stability!

Jim Anchin, CEO of the accounting firm of Anchin, Block & Anchin, LLP, and construction industry specialist, believes there are two major reasons why contractor clients tend to shy away from the numbers: They wear too many hats and don't take enough time for important administrative tasks, even if it could save them money; they aren't comfortable with the details of accounting and financial management. Tom Rogers, senior construction lender at Signature Bank, finds that contractors "often lack understanding of the financial consequences of their work and the impact on financials."

In all fairness, you probably encounter no greater challenge as a contractor than having to manage your financial picture successfully. To win business, you must compete based on price, which creates a tendency to shave your bid at the expense of your profit margin. Once you're on the job, predicting and tracking cash flows and profitability are difficult because you're not entirely in control of events. Your project payment stream can be derailed because of delays in customer progress payments and change order approvals, schedule changes, or disagreements regarding scope of work. Your profitability can be impaired because of rejected change order requests, unanticipated labor and materials costs, extra costs incurred in correcting defective work, and difficulties in obtaining full payment for "retainage" (or retention). These problems lead to cash-flow crunches, "profit fade," and even losses, which, in turn, eat into your equity base.

If anyone has told you that you need deep pockets to succeed in the construction industry, they're speaking the truth. Because of the uncertainties inherent in your revenues, cash flows, and costs, you must build an equity base that can sustain you through the ups and downs and keep you ahead of that "desperation curve." Dennis J. Chamberlain, regional vice president of St. Paul Travelers, a leading nationwide insurance company, observed that "Contractors often find out through slow receivable turnover and challenging projects just how important it is to grow and maintain capital in the business. This provides staying power to face cash-flow and project challenges."

To build your equity base, you must obtain a fair and reasonable price for your work (see "Creating a Smart Bidding Strategy" in Chapter 7), build a cost-effective operation, maximize cash flows, monitor your overall financial performance, secure outside sources of capital and support, and reinvest your profits in the company.

That's a tall order that requires you to have a grip on accounting and financial principles and take the time to actively manage your financial situation. If this isn't one of your strengths, make it a priority to learn all you can about accounting and financial management. Take courses at your local college, read a good construction accounting and finance book and keep it handy as a reference, and take advantage of seminars and other resources offered by industry associations.

In addition, lean on your construction accountant for expertise and advice and hire a construction-savvy bookkeeper, controller, or financial manager as soon as you can afford to. Lender Tom Rogers concludes, "You want the contractor to have a strong internal accounting system to monitor job performance and a CPA firm with experience in the construction field."

BUDGETING AND CONTROLLING COMPANY COSTS

A successful specialty contractor who has been in business for more than 30 years said it eloquently: "If you don't know your costs, you're finished." With the right job-costing and financial-reporting procedures and systems in place, you should be able to rely on accurate current and historical financial data from your internal bookkeeping system to create company budgets and forecasts, review costs, and manage profitability.

Armed with the facts, you can prepare monthly and annual *budgets* that will serve as your tools to measure company progress relative to goals. Using your income statement format as a template for your budget, you can start off with an estimate of your revenues, costs, and profits for 12 months based on the following:

- Project revenues and direct costs will be derived from individual project budgets for all work in progress and from bid and schedule information for your new work backlog (awarded and signed contracts).
- Company overhead costs will be extrapolated from historical experience.
- Your profit margin should represent a realistic goal given your past performance and assessment of market trends.

By comparing monthly and year-to-date actual numbers with your budget, you can determine where you need to take corrective action. Going down the items in your income statement, here are some key cost components that should be actively managed:

Project Costs
- *Materials.* You can prepurchase certain materials you believe will become more expensive during the course of your work. As long as you take possession of the materials and store them safely, you'll be able to include them in your payment requisition. You can also make bulk purchases for two or more concurrent jobs so that you take advantage of volume discounts. In addition, you can

SAMPLE MONTHLY BUDGET COMPARISON[1]

Year to Date	Budget	Actual	Variance
REVENUES	$	$	$
DIRECT PROJECT COSTS			
Materials			
Labor			
Subcontracts			
Supervision			
Bonding, Insurance			
Permits, Licenses, Fees			
Equipment			
DIRECT PROJECT COSTS			
GROSS PROFIT*			
OPERATING EXPENSES			
Variable Costs			
Advertising			
Communications			
Interest			
Office Supplies			
Taxes			
Travel and Entertainment			
Fixed Costs			
Depreciation			
Dues			
Professional Fees			
Rent			
Repairs and Maintenance			
Officer and Office Salaries			
Supplies and Tools			
Taxes			
Utilities			
TOTAL OPERATING EXPENSES			
NET PROFIT BEFORE TAX[†]			

* *Revenues Minus Direct Project Costs*
†*Gross Profit Minus Operating Expenses*

minimize extra material costs resulting from corrective work by having good quality controls in place. Above all, make sure that your contract allows you to pass through inflationary price increases in materials and labor costs.

- *Labor.* Your manpower costs could escalate well beyond what you budgeted because of poor labor productivity. Therefore, it's critical that you determine daily production goals and monitor actual performance before the situation gets out of control. The best antidote to poor worker performance is a good field superintendent who communicates plans daily, coordinates the work, and makes sure that materials, tools, and equipment are available and in good working order.

- *Equipment.* The number of days a piece of equipment is assigned to a project and the hours of operation must be tracked so you can move equipment efficiently from one project to the other. A good maintenance and repair plan will also help reduce idle time.

- *Subcontractor buyouts.* Finalize your purchasing as quickly as possible to minimize the risk of price escalation. Put another way, if you leave line-item costs as "uncommitted," you run the risk of having to incur additional costs when you finally buy out the item.

- *Change orders.* Efficient procedures will ensure that change order requests are approved quickly. Outstanding unapproved change order proposals represent monies that you may never receive for work that you may have already performed.

- *Site supervision.* The size of your site staff must match your current and future work commitments. When you're busy, not having the right quality or amount of site supervision can lead to extra costs because of mistakes, lack of follow-up, poor labor productivity, and disputes, which come right out of your bottom line. If your project backlog (awarded and signed contracts for new projects) is weaker than expected, you may need to reduce the size of your staff and do so in a timely fashion to minimize the drag on profits.

- *Insurance.* Your insurance costs are based on your safety record as captured by your "experience modification rating" or EMR, which increases or decreases depending on the number of incidents and accidents reported on your job sites. Strict adherence to your safety program will keep insurance costs in check.

Company Operating Expenses

These general and administrative company expenses represent your overhead costs that are distributed among all your projects. Review each cost and determine whether and how you can reduce it:

- *Variable overhead* that can be reduced includes *interest expense* (by minimizing borrowings), *office supplies* (by purchasing periodically and receiving bulk discounts), *telephone/Internet communications* (by negotiating a better deal); *advertising* (by evaluating the effectiveness of your effort), and *travel and entertainment* (by setting a reasonable budget and staying with it).
- *Fixed overhead* is more difficult to control because it includes items such as rent, utilities, and professional fees (legal and accounting). The largest item is the one that can be reduced—*officer and office salaries* that are not direct project costs. In slow periods, you always have the option to reduce your compensation. You can also shrink your office staff as long as you don't compromise your company's ability to function effectively.

MANAGING CASH FLOWS

You've probably heard the phrase "Cash is king" umpteen times because it's true; without having cash to pay your bills, you no longer have a business. Therefore, to avoid unpleasant surprises, you must understand how money flows through your business and how to anticipate cash needs. As one contractor stressed above all else, "Make sure you have cash!"

Good cash management is a special challenge for construction contractors because of the uncertainties inherent in the business; you must provide a hard price at bid before your final costs have been determined, and you have to wait to get paid for work you already completed, resulting in gaps in timing between expenditures and receipts. The challenge can present serious difficulties if you don't have sufficient cash or bank credit to fill in those gaps.

To avoid a deadly cash crunch, you must make every effort to *maximize and accelerate* the amount of cash flowing into your account, *control* the amount of cash flowing out, and *forecast* your cash needs so that you can cover expected shortfalls and plan for necessary investments. Following are the methods that successful contractors typically use to manage their cash flows.

Maximize and Accelerate Cash Inflow

- Before you sign a new contract, try to negotiate fair and reasonable payment terms and a retainage (or retention) amount.
- To jump-start project cash flows, get your schedule of values approved as soon as you sign your contract and submit your first requisition shortly thereafter. Your invoice should include mobilization costs (i.e., site office setup, supervision, temporary facilities) and bonding and insurance costs. This will help minimize the use of internal cash or borrowings to get the project off the ground.
- When you negotiate your schedule of values with your customer, assign a slightly higher percentage of the contract price to early work items. A reasonable amount of "front-end loading" allows you to bill and collect amounts in excess of costs incurred, which accelerates cash flow early on in the project. This practice is effective if the cash generated remains available to cover cash shortfalls later on. More often, contractors tend to move money from a project that has a positive cash flow to one that needs a cash injection. "Borrowing from Peter to pay Paul" is a sure sign that trouble is brewing.
- Generate accurate and complete payment requisitions promptly and get them to the customer as quickly as possible. Institute tight procedures to track your receivables aging and follow up on collections. Try not to hold up payment unnecessarily because of minor issues.
- Practice prudent contract and change order management. Referring to the essence of Chapter 7, you'll greatly improve your chances of getting paid in full and in a timely fashion if you perform the work in accordance with your contract and submit proper and timely documentation.

- Manage your schedule properly—the faster you finish the job, the better your cash flow.
- During the punch-list phase, do all you can to accelerate resolution of open items and final acceptance by your customer so you can receive final payment, including the retention amount.
- Minimize idle cash in your bank accounts by keeping it in interest-bearing accounts or by paying down loans to reduce interest expense. Ask your customers to send payments by electronic transfer.

Control Cash Outflow

- Scrutinize every expense periodically, both out in the field and in the office, to identify opportunities for cost savings, as discussed earlier in this chapter.
- Manage your work-in-progress billings so that *overbillings* (billings for work not yet performed, a liability) is greater than *underbillings* (unbilled receivables, an asset), meaning that your work is being funded more by the customer than by internal cash.[2]
- Utilize your cash-flow projections to make decisions regarding the timing of necessary expenditures (i.e., office and field equipment, software, office space).
- Retain at least the same amount from your subcontractors as your customer retains from you.
- Make sure you maintain a good credit rating so you can negotiate favorable terms with your vendors and suppliers. Schedule payments just before their due dates.
- Don't delay payroll and income tax payments. The longer you wait, the deeper in the hole you'll be and the higher the penalties you'll pay, not to mention the fact that you could get into serious trouble with the U.S. Department of Labor and the Internal Revenue Service!

Forecast Your Cash Needs

- Your monthly cash-flow forecast will map out cash inflows (progress payments due), cash outflows (payments to subcontractors, vendors, and suppliers and other expenses), and track cash surpluses and deficits for each period.

- Before you mobilize on a project, prepare a project cash-flow forecast based on your schedule and work breakdown. Map out anticipated progress payments and expenditures so that you can determine if and when additional cash resources will be needed on the project.
- Your cash-flow forecast for each project should be incorporated into your total company cash-flow forecast. Use this information to gauge performance, take corrective action where you can, and communicate with your banker about the nature and extent of your borrowing needs.

SAMPLE PROJECT CASH-FLOW FORECAST

	Month 1	Month 2	Month 3	Month 4 ...	TOTAL
BILLINGS					
Progress Payments					
Less Retainage					
Retainage Paid					
Change Orders					
TOTAL CASH IN					
CASH OUT					
Materials/Supplies					
Subcontractors					
Labor					
Other Direct Costs					
TOTAL CASH OUT					
NET CASH FLOW*					
CUMULATIVE CASH FLOW†					

*Cash In Minus Cash Out
†Cumulative Cash Flow Month 1 Plus Net Cash Flow Month 2

SAMPLE COMPANY CASH-FLOW FORECAST[3]

	Month 1	Month 2	Month 3	Month 4 . . .	TOTAL
CASH BALANCE					
CASH IN					
Job 1					
Job 2					
Job 3					
TOTAL CASH IN					
CASH OUT					
Job 1					
Job 2					
Job 3					
Taxes					
Equipment					
Insurance, Bonding					
Overhead					
Direct Labor					
Other Direct Costs					
TOTAL CASH OUT					
NET CASH FLOW*					
NEW BORROWINGS†					
LOAN REPAYMENTS‡					
CASH BALANCE§					

*Cash Balance Plus Cash In Minus Cash Out
†Cash Shortfall
‡Cash Surplus
§Cash Balance Plus or Minus New Borrowings/Loan Repayments

MONITORING YOUR FINANCIAL PERFORMANCE

Every year, your accountant will prepare formal financial statements in addition to your tax returns so that you'll meet the reporting requirements of your bank and bonding company. Your financials will include an income statement, a sources and uses of funds statement, a balance sheet, and the accountant's notes and schedules. The *income statement* reflects sales, direct costs, overhead, and profit or loss for a specific period of time. The *balance sheet* is a snapshot of your company's financial position (assets, obligations, equity) at a specific point in time. The *sources and uses of funds statement* links the income statement to the balance sheet.

Throughout the year, you'll also be generating financial information directly from your internal bookkeeping system whenever you need it, including income statements, cash-flow statements, balance sheets, receivables and payable aging schedules, and individual job costs. Your data is raw and more detailed than your accountant's reports and will provide you with a snapshot of where you are at any point in time. Of course, the more accurate your records are, the clearer the snapshot.

The reports you generate are your primary tools for planning, coordinating, and controlling your company's activities and performance. At a minimum, you should keep track of your cash position, receivables, individual project costs, and total work-in-progress (WIP) information on a frequent basis and review your monthly income statements as soon as they are available. Being up-to-date allows you to determine percentage completion, costs-to-complete, and profit margins on each project and actively manage your company's cash position and profit performance.

Going beyond the routine, you should review your financial performance periodically to gauge the trends: Are your profit margins improving or deteriorating? Is your working capital growing or shrinking? Are your receivables under control? Is your debt at a manageable level? How are you doing overall compared to last period, last year? If you've been in business for a few years, how are you doing compared to similar companies in the industry? Your analysis will allow you to compare actual with anticipated performance, identify areas of weakness, and take action to improve results before it's too late.

SAMPLE SUMMARY BALANCE SHEET

ASSETS

    Current Assets

    – *Cash* $_____

    – *Accounts Receivable*

    – *Materials Inventory*

    – *Notes Receivable*

    – *Underbillings**

    – *Prepaid Expenses*

    – *Other Current Assets*

    Total Current Assets

    Fixed Assets

    – *Less Accumulated Depreciation*

    Net Fixed Assets

    Other Assets

TOTAL ASSETS

LIABILITIES

    Current Liabilities

    – *Accounts Payable*

    – *Notes Payable*

    – *Overbillings†*

    – *Accrued Expenses*

    – *Other Current Liabilities*

    Total Current Liabilities

    Long-Term Liabilities

TOTAL LIABILITIES

OWNER'S EQUITY

    Capital Stock

    Retained Earnings

TOTAL OWNER'S EQUITY

TOTAL LIABILITIES AND OWNER'S EQUITY

*Costs and Estimated Earnings in Excess of Billings
†Billings in Excess of Costs and Estimated Earnings

To enhance your analysis and diagnosis, you can use *financial ratios* that are commonly used by your banker and your bonding company to obtain indications of your performance. These key ratios can help you monitor trends in your profitability, liquidity, and capitalization and compare your performance with similar construction companies in your geographic region.

To Track Profitability

- Your *gross profit margin* (your gross profit divided by revenue) reflects how well you're pricing and managing costs on your projects. An increasing gross profit margin indicates improvements in pricing and/or job cost controls.
- Your *overhead rate* (overhead expenses divided by revenue) confirms whether company overhead is increasing or decreasing relative to revenues and what percentage you must add to your cost estimate to recover your operating expenses. An increasing overhead rate indicates that either revenues are falling or overhead costs are increasing, or both.
- Your *net margin* (net income divided by revenue) measures your performance managing job costs and company overhead.
- Your *return on assets* (net profit divided by fixed assets) indicates the profit generated by your fixed assets. The higher the ratio, the better utilized your assets are.
- Your *return on equity* (net income divided by net worth) tells you whether you're earning an acceptable return given the risks you take in the business.

To Track Liquidity

- Your *current ratio* (current assets divided by current liabilities) measures the extent to which your short-term obligations are covered by current assets that can be converted into cash. The higher the current ratio, the easier it is for you to meet your immediate obligations and generate cash to invest in your business.
- Your *quick ratio* (current assets minus inventory divided by current liabilities) measures how well you can pay short-term liabilities without relying on the sale of materials inventory.

- Your *working capital* (current assets minus current liabilities) measures the excess cash you generate after taking care of all short-term obligations.
- Your *current assets to total assets* indicates how much of your total assets can be converted to cash quickly. If the ratio is low, it means that your fixed assets are high or that you have little cash. A high ratio means that you have too many receivables outstanding or too much idle cash.
- The *average age of your accounts receivable* tells you how long your receivables remain outstanding. If it's more than 45 days, this means that too much working capital is tied up and isn't being used to generate more income.
- The *average age of accounts payable* tells you how long your payables remain outstanding. A number above 45 days indicates that you're a slow payer and are not taking advantage of trade discounts.

To Track Capitalization

- Your *liabilities to equity ratio* measures how balanced your capitalization is. A high ratio means that you have too much debt and may not be able to repay interest and principal. A low ratio means that you may not be taking sufficient advantage of financing opportunities to grow your business.

For your convenience, the chart that follows lists key financial ratios and some industry benchmarks.[4] Data for your specific trade or specialty, company size, and geographic region can be obtained from such organizations as Robert Morris Associates *(Annual Statement Studies)*, Standard & Poor's, the National Association of Home Builders *(Cost of Doing Business* study) and the Construction Financial Management Association. You can find much of this information in the reference section of most business libraries.

MEETING YOUR FINANCING NEEDS

I mentioned in Chapter 2 that bankers in general don't get a "warm and fuzzy feeling" when you first walk in the door because they don't understand the nature of the construction contracting business and may

SOME INDUSTRY BENCHMARKS[5]

Key Ratios	Calculation	Recommended Range
Profitability		
– Gross Margin	$\dfrac{Revenue - Direct\ Project\ Costs}{Revenue}$	N/A
– Overhead Rate	$\dfrac{Company\ Overhead}{Revenue}$	N/A
	$\dfrac{Company\ Overhead}{Direct\ Project\ Costs}$	N/A
– Net Margin	$\dfrac{Net\ Profit\ Before\ Taxes}{Revenue}$	Min. 5%
– Return on Equity	$\dfrac{Net\ Profit\ Before\ Taxes}{Net\ Worth}$	Min. 15%
Liquidity/Turnover		
– Current Ratio	$\dfrac{Current\ Assets}{Current\ Liabilities}$	1.5 – 2.0 to 1
– Quick Ratio	$\dfrac{Cash + Accts.\ Receivable}{Current\ Liabilities}$	1.0 – 1.5 to 1
– Current Assets/ Total Assets	$\dfrac{Current\ Assets}{Total\ Assets}$	0.60 – 0.80 to 1
– Accts. Receivable Age	$\dfrac{Accts.\ Receivable \times 365}{Revenue}$	Max. 45 days
– Accts. Payable Age	$\dfrac{Accts.\ Payable \times 365}{Direct\ Project\ Costs}$	Max. 45 days
Capitalization/Leverage		
– Total Liabilities to Equity	$\dfrac{Total\ Liabilities}{Net\ Worth}$	1.0 – 2.0 to 1

have preconceived notions about the industry. In fact, many banks shy away from lending to contractors altogether unless the companies are very large and well established. Why? Because, going back to Chapter 1, construction is a risky business with high rates of business failure and bankruptcy.

The lenders who do extend credit to small and medium-size construction companies are typically regional and local banks in your geographic area. Their construction lending experts are knowledgeable about local economic and political conditions and are well acquainted with local experts and trade associations. Ask your accountant, attorney, bonding agent, or consultant to make an introduction to those lenders in your area.

Obtaining your first line of credit, even only for a small amount, will help you get in the door and develop a good banking relationship. If you've been in business for a year or two and are shopping around for a line of credit for the first time, find out which banks offer a "no-doc" or "low-doc" program. These lines of credit are approved based on limited information (i.e., your tax returns, credit history, and personal guarantees) and, in many cases, are backed by a guarantee from the U.S. Small Business Administration for up to 90 percent of the loan. For a traditional line of credit, you'll have to provide detailed documentation, including company history, financial statements, work-in-progress information, and customer and vendor references. Given that your line of credit will most likely be unsecured, lenders will scrutinize your personal background, experience and track record, company profit history and financial ratios, and the amount of equity you have in the business. They'll look at your cash-flow generating capabilities, the quality of your receivables, and the size of your personal assets to confirm that your loan can be repaid. They'll likely require a personal guarantee from you and your partners.

As your company grows and your working capital needs expand, you'll need to increase your line of credit. For larger loan amounts, the bank will require that you submit recent financial statements, work-in-progress schedules, and project-specific cost and profitability information on a regular basis. You'll also need to keep your loan officer abreast of all company developments—good and bad—if you intend to gain and maintain trust and confidence. You'll quickly lose their support if you over-promise on performance or neglect to communicate a problem that could impact your financial position. In a nutshell: bankers don't like surprises.

You'll gain more flexibility as you build your financial strength and solidify your banking relationship. Signed construction contracts and well-developed project-cost and cash-flow projections, for instance, could help you obtain an increase in your line of credit based on anticipated revenues. Superior liquidity and capitalization could eventually convince your lender to remove personal guarantees attached to your line of credit.

SATISFYING BONDING REQUIREMENTS

If you're a general contractor who's interested in doing work for federal, state, or local governments, you'll be required to provide bonding to guarantee your work. The Miller Act of 1935 requires that all federal projects over $100,000 are bonded by a surety, and each state has similar laws. If you're a subcontractor, the prime contractor may require that you provide a bond. Many private owners and developers require a bond as well.

The purpose of the bond is to protect the public, subcontractors, and vendors by shifting the risks of project completion from the project owner to the surety company. The *bid bond* guarantees that you'll honor your price; the *payment bond* guarantees payment to your subcontractors, laborers, and suppliers; and the *performance bond* protects the project owner from financial loss if you fail to perform the contract in accordance with terms and conditions.

By issuing a "bond line" on your behalf, a surety company in essence *prequalifies* your company to do the work based on your qualifications and financial strength. If you perform well and establish a good track record, the surety will likely increase your bonding capacity, allowing you to bid on more and bigger jobs. If you're unable to perform, the surety will declare you in default and step in to complete the project. It will charge you for losses, expenses, and attorney's fees and terminate your bond line.

No matter where you're located geographically, qualifying for bonding is a tough challenge these days. After suffering huge losses in recent years,y mainly because of aggressive expansion, the surety industry has retreated significantly. Some have reduced their surety operations while others have withdrawn from the market altogether. Those that are still in the market have tightened their underwriting requirements and have become more selective in efforts to bring losses to a minimum. Hank Harris, president of FMI Corporation, a management consultancy specializing in the construction industry, has seen more joint ventures and mergers recently among construction firms as a result of these conditions.

In the face of this difficult environment, how can you qualify for bonding? I talked with several bonding agents and surety professionals and they relayed the following tips:

- The quality of your financial statements is very important to a surety; numbers are your ultimate scorecard. Sureties prefer that you have them prepared by a construction accountant.
- As CEO of your company, make sure that you're able to analyze and present your financial results yourself. Your knowledge and ability to articulate your company's situation will go a long way toward establishing credibility.
- Surety companies usually distribute bonds through independent agents or brokers, also known as "surety bond producers." You should choose your agent carefully because he or she will be presenting your company to sureties. A good bonding agent could prove to be invaluable in providing business and technical advice and introducing you to accountants, bankers, lawyers, and other professionals who serve the construction industry. Your accountant will probably know a good agent in your geographic area. You can also locate an agent by visiting the Web site of the National Association of Surety Bond Producers at *www.nasbp.org.*
- Keep informed about current surety industry trends and learn about the underwriting process. Know what the terms and conditions of your bond are—*understand* what you're getting yourself into.
- Rely on your bonding agent to contact the appropriate sureties for underwriting. He or she will need the following information to support your case:
  - The history of your company
  - An organization chart and detailed résumés of key personnel
  - A description of project experience with customer recommendation letters and references from subcontractors and suppliers, and details of your largest completed job to date

- Specific information on all projects in progress, including project billings and costs to date, revised contract amounts, costs to complete, and percentage of completion
- At least three years of company and personal financial statements
- A company business plan and a continuity plan (The surety wants to see that you have made provisions for your succession via a "buy-sell agreement" and "key man" life insurance.)
- A bank line of credit that covers your working capital needs
- The bonding decision will be based on several factors, including your character (i.e., honesty, integrity, commitment, professionalism) and experience, your capacity to perform (i.e., personnel, equipment, expertise, relationships with subcontractors and vendors), your financial strength relative to the construction contract amount, your credit history and financial ratios, and available lines of credit.
- According to Bill Haas, senior bond producer with USI Holdings Corporation, a national insurance brokerage firm, the amount of your bond line can be determined in several ways, including a multiple of the dollar value of your largest completed project (the current benchmark is around 2 times) or a multiple of your working capital (ranging from 10 to 20 times depending on your track record). Haas noted that sureties typically exclude all receivables 90+ days past due from your working capital number.
- You'll be paying a fee of between 1 percent and 3 percent of the contract price for the bond, to be included in your project direct costs.

As I described earlier, it's increasingly difficult to obtain bonding, especially the first time. Scott Adams, president of Avalon Risk, LLC, a general agency specializing in surety bonds for small contractors, advises that you "diversify your client base between public- and private-sector work so you're not held hostage to bonded work, and therefore to the underwriting cycles of the bonding business."

If you're a minority-owned, woman-owned, small, or disadvantaged business, look for special government agency programs that provide bonding assistance or that exempt you from the bonding requirement up to a certain dollar amount.

Sources of Bonding Information[6]

The **U.S. Small Business Administration** under its SBA Surety Bond Guarantee Program at *www.sba.gov/OSG* or by calling (202) 205-6540

The **National Association of Surety Bond Producers** (NASBP) at *www.nasbp.org* or at (202) 686-3700 to find a bonding agent in your area

The **Surety Information Office** (SIO) at *www.sio.org* or call (202) 686-7563 for information on how to obtain a bond

The **Surety Association of America** (SAA) at *www.surety.org* for general trends and issues

BUILDING YOUR EQUITY BASE

It's common knowledge among contractors that the biggest reason for business failure in the construction industry is undercapitalization. With low profit margins and relatively high business risks, it's no easy task to accumulate equity and establish financial stability.

So far in this chapter, we've explored how you can maximize your bottom line by obtaining a fair price for your work, controlling company costs, managing cash flows, and closely monitoring your financial performance. With reasonable profits, cash balances, and liquidity, you can reinvest in your business and accumulate equity. You'll gain comfort from knowing that you don't have to take every job that comes along and that you have a cushion against slow periods and sudden setbacks.

To build an equity base, however, you must keep your profits in the business. Peter Lehrer, co-founder of Lehrer McGovern, advises, "Above all, make a long-term investment. It's your responsibility to your customers, people, and suppliers." Put bluntly, resist the urge to take money out of the company for personal use. In Jim Anchin's experience as accountant to the construction industry, he says, "Contractors tend to overcompensate themselves personally when cash balances are up, leaving the company to suffer when cash balances are down or conditions change."

It's also about making sure not to subject your company to unnecessary risks and expenditures. Scott Adams of Avalon Risk cautions his clients to only take risks that match their level of capitalization.

You'll be putting your company at peril if you overreach your capabilities by taking on contracts that greatly exceed your equity or by trying to grow your business infrastructure without adequate financial resources.

Similarly, resist the temptation to use company funds to dabble in unrelated businesses, such as real estate or restaurants. Eric Kreuter, senior partner of the construction accounting firm of Marden, Harrison, Kreuter, CPAs, P.C., has seen it all too many times: "Contractors who decide to dabble in side projects fragment their capabilities and subject their companies to unnecessary risks."

Now You Have It, Now You Don't

XYZ Specialty Company, a 15-year-old specialty contractor with a history of success and profits doing work for public- and private-sector customers, is owned by two brothers. One bids on the work and manages the business while the other focuses on the details of fabrication and installation. In the mid-1990s, they hit their stride and executed a steady stream of profitable projects. They delivered good work and their customers were happy. By 2001, XYZ had spacious offices, fancy company cars for each member of the management team, and several million dollars in the bank. The brothers were feeling so confident and prosperous that they decided to diversify their business by investing in a couple of unrelated ventures. They plunged into their new activities with great enthusiasm and energy and left the core business in the hands of their field managers. Work performance started to drop, billing and receivables collection deteriorated, and several jobs were poorly bid. In the meantime, their new ventures had not taken off as expected. By 2004, the company's cash cushion had whittled down close to zero and several key people had jumped ship. Today, the brothers are scrambling to deal with customer complaints, uncollected receivables, unpaid bills, and creditors breathing down their backs.

Lessons Learned: The brothers plunged company resources into risky business ventures that haven't yielded promised returns. In the meantime, their company infrastructure (people, organization, systems) has not been adequate to support their expanded activities. The core business has suffered, probably beyond repair.

More Reading on Financial Management

Jackson, Jerry. *Financial Management for Contractors.* FMI Corporation, 2002.

Peterson, Steven J. *Construction Accounting and Financial Management.* Prentice Hall, 2004.

10

RISE TO THE CHALLENGE IN A CHANGING ENVIRONMENT

Conditions in the construction market are always changing; construction activity fluctuates up and down mainly as a result of economic forces. Specific types of construction are in or out of vogue depending on demand. Last year's sizzling residential market will perhaps be replaced by this year's new "hot" markets in, for example, office, retail, and mixed-use developments. Some regional markets that are "booming" now may be fading before too long. These constant shifts have a significant impact on construction companies as they scramble to adapt to the latest market trends and avoid being left out of the game.

The construction industry itself is in a state of transition as it tries to keep up with an increasingly complex and demanding world. Project owners' expectations are higher than ever with regard to contractor expertise, quality of work, service, and ability to meet tight budget and schedule requirements. Owners are also as eager as ever to shift the bulk of the risk to the constructor. The work itself is becoming more complex as better, smarter, and "greener" building designs and systems are becoming popular. New project delivery alternatives are sprouting in every direction in efforts by construction companies to offer something "new" and to satisfy increasingly sophisticated customer and project needs.

And yet, some things never change. We're still faced with age-old industry weaknesses and problems with no resolution in sight, including low profitability, intense price competition, material price fluctuations, inefficient purchasing methods, shortages in labor and materials, combative relationships on the job site, and legal disputes. Compared with other major industries, we've made limited progress in embracing technology, don't devote much effort to real innovation, and have done little to develop a pool of good management talent. And, to top it off, we continue to be undercapitalized and vulnerable to the changing appetites of the bank lending and surety markets.

The contractors who hold their own in this environment tend to be the niche players or the industry giants. The *niche players* are small and midsize companies that focus on providing particular expertise and a high level of service to their customers. One example that comes to mind is a highly successful midsize general contractor known specifically for executing beautiful high-end work on signature design buildings, only in Manhattan. The *industry giants* have the capacity and resources to deliver a complete menu of services, from soup to nuts, and to assume the financial risk on large construction projects. These companies tend to dominate large construction in regional and national markets.

The vast majority of contractors are established businesses that don't have a strong identity or financial wherewithal. They're set in their ways, aren't as plugged into industry and market trends as they should be, and struggle to live in crisis management and survival modes year in and year out.

As a business owner of a start-up or growing construction company, you're fully aware that it's not likely you're going to be an industry leader overnight. You're also set on not being lumped in with the "majority of contractors" who struggle to keep their heads above water.

As a growing construction company, how do you gain a competitive edge so you're not left in the dust? How do you organize your business so it can thrive in a constantly changing market environment? *What must you do to make sure you succeed?*

Based on my professional experience and the feedback I've received from a wide variety of construction company CEOs and industry experts, I believe that there are three key drivers (as shown in Figure 10.1) to gaining a competitive edge: *people, system, and service.*

- **People.**

  To execute a construction project, you must have people, so they're by far your most important assets. If you hire and retain the best people, you gain a competitive edge and success follows.

- **System.**

  To deliver the best possible price, you must develop an efficient and cost-effective organization that enables people to be highly productive.

- **Service.**

  To win customer satisfaction and repeat business, you must build a company culture that revolves around customer service.

The more competent and efficient your people and organization are, respectively, and the more focused your company is on customer service, the stronger your competitive advantage.

Terry L. Yeager, an associate director of Navigant Consulting, an international consulting firm, puts it another way: "To be successful, construction companies must become market-based, customer-focused, and performance-driven. . . . A construction company must optimize the opportunities in the market while minimizing the internal costs to deliver the service or product while making certain the customer is satisfied with the services rendered. Management must provide momentum and direction . . . and constantly find ways to integrate and sharpen efforts to satisfy the customer's demands, drive out all unnecessary costs, and maximize the profitability of the company."

FIGURE 10.1 *Gaining a Competitive Edge*

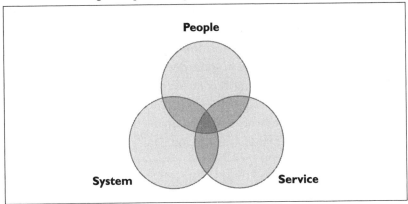

One piece of good news is that you can gain that competitive edge by starting off already "market-based, customer-focused, and performance-driven." Unlike some of the big players, who are burdened by large organizations, high overhead costs, and entrenched cultures, you're a "work in progress," capable of being nimble, creative, dynamic, and innovative. According to management guru Peter F. Drucker, "The entrepreneur always searches for change, responds to it, and exploits it as an opportunity."

Wearing your entrepreneurial hat, and keeping the three key drivers to competitive advantage in mind, you can lay the foundation for stability and success by organizing your company properly from the start:

- Instead of "playing it by ear" as is common in construction, you can adopt strategic planning as a company and project management tool and make it part of your company culture.
- Armed with access to almost limitless information about industry and market trends, you can be smart about positioning your company and developing services that directly address customers' needs.
- Instead of scrambling to hire personnel to meet immediate needs, you can plan to build a top core management team by systematically looking within and outside of the industry for talent and by designing better ways to train, motivate, and reward your employees.
- You can also choose to embrace new systems and technologies that will help you create a cost-effective and competitive operation.

The other piece of good news I've tried to convey in this book is this: it's entirely possible to *insulate* your company from market and industry ebbs and flows and lay the foundation for long-term success. In Chapters 1 through 9, we reviewed in some detail the key ingredients to business success: understanding business risks and challenges; careful planning; targeted marketing; effective company organization; strong in-house team and outside resources; company "best practices"; effective project management; prudent protection against risk; proper financial systems and controls.

Hopefully, you've also taken advantage of the tools I've made available to help you navigate through the rough waters of setting up a successful business, including the Self-Assessment Questionnaire, the Goals Worksheet, the Strategic Business Planning Workbook, the Market Positioning Statement, the Bid Decision, and the Project Management Procedures Workbook. They were designed to help you develop a real appreciation for what it takes and to start laying out a workable plan to achieve your goals. Many of these tools are available in the Appendixes and online at *www.ConstructBiz.com*.

Before I wrap up, I'd like to review the key ingredients covered in this book:

- **Capability and commitment.**

  Make sure you have what it takes. Take a good hard look at yourself and confirm that you've got the personality, inclination, experience, capabilities, resources, and commitment to run a construction company. Some key questions are: Do I have the right experience and expertise? Do I have "passion and focus"? Am I self-disciplined and organized? Am I a self-starter? Do I have the stomach to deal with uncertainties and risks? Do I have sufficient business experience and financial resources? Do I have a good support network? If you're weak in one or more of these, consider going back to work for someone else.

- **Understanding.**

  Know what's going on. It's critical that you are plugged into the latest news and gain a thorough knowledge of market and industry conditions and trends. As I suggested in Chapter 2, you should be asking yourself constantly, "What are the general trends and where are they heading? How can I anticipate the market and ensure that my business is a leader/trendsetter and not a follower? How can I prepare my company against threats I see looming in the horizon?"

- **Plans.**

  Create a decision-making framework. Answer the basic strategic question: "Where do I want to go with my life, with my business?" Define your business and market and plan your goals and approach accordingly. Create realistic goals and objectives and include your team in the implementation process. Identify obstacles to success and how you can overcome them. Compare actual performance with expectations and adjust priorities and plans. Always strive for improvement.

- **Message.**

  Deliver a clear message about your advantages. Based on the results of your research, define your market and position your company so that it will be noticed and eventually selected. Create a clear message that conveys your customer value offer and your company culture (i.e., integrity, professionalism, commitment). Establish that you're unique, different, better.

- **Customers.**

  Understand what your target customers need. Do the research and ask the questions; tailor your services accordingly and obtain customer feedback on your performance. Make it a top priority to provide excellent customer service and follow-up and train and motivate your staff to deliver every time.

- **Leadership.**

  Provide direction and a positive outlook. Share your plans and goals clearly and provide the tools your team will need to implement them. Set priorities and empower your people to take responsibility. Create a system that builds competence and capability. Foster enthusiasm, discipline, trust, respect, open communication, and collaboration.

- **Organization.**

  Build an organization that delivers. Develop the capability to perform by setting up a basic organizational structure that can evolve as your business grows. Integrate office and field operations so they work smarter individually and collectively. Surround your company with resources and support.

- **People.**

  Build a versatile team that can adapt to changing conditions. Make it your top priority to bring in and retain talented, competent, and dedicated people who will become part of your core team. Provide training, fair compensation, and a sense of accountability and belonging. Treat your people with respect and dignity.

- **Communication.**

  Maximize company efficiency. Recognize that the services your company provides are time-sensitive and paper-intensive. Manage information "overload" by creating standard practices and procedures, developing a system of reporting and communication, and taking advantage of technology to promote worker productivity and collaboration.

- **Management and controls.**
  Deliver what you promised. Make sure you have strong in-house capabilities in estimating, project management, and cost controls. Establish a clear-cut approach and methodology for planning and managing construction projects. Ensure that you have a qualified project management team and reliable subcontractors, vendors, and suppliers. Review performance and learn from mistakes.

- **Protection.**
  Minimize exposure to liability. Know what your rights and obligations are under your construction contracts and actively manage your projects to avoid disputes. Establish a risk management strategy. Make safety a top priority and minimize the risk of loss and injury at the job site.

- **Stability.**
  Maximize profits and build a cash cushion. Get a grip on accounting and financial principles and actively manage your financial situation. Focus on budgeting and controlling company costs to maximize profits. Closely manage your cash flows. Leverage your capabilities with bank and surety support. Build your equity base by reinvesting in your business.

As you embark on your exciting adventure of starting and growing a construction business, the trick is to include all these elements in your thinking. Like it or not, your business is like a house of cards; remove one of the cards and your house becomes shaky and eventually falls flat. With your dedication and enthusiasm, and with the support of your partners, employees, advisors, family, and friends, you'll be able to map out your own path to success that works. I hope this book will prove helpful as a compass, resource, and source of moral support.

Chapter 1

1. Ken Simonson, "Quick Facts about the Construction Industry" (January 17, 2006).

2. U.S. Census Bureau, *Construction Spending Report* (November 2005). *www.census.gov/const/www/c30index.html*

3. Simonson.

4. Michael Porter, *Competitive Strategy: Techniques for Analyzing Industries and Competitors* (Free Press, 1998).

5. Simonson.

6. Ibid.

7. Sue Kirchhoff, "Labor Study Measures Rising Building Costs," *USA Today* (January 30, 2006).

Chapter 6

1. Tom Sawyer, "Online Management Tools Excel at Empowering Project Teams," *Engineering News Record* (October 11, 2004).

Chapter 8

1. Ira M. Schulman and Thomas H. Welby, "Construction Law: Can This Job Be Saved?" Presentation. February 9, 2000. Tarrytown, New York.

2. AIA Document A201—Agreement: Contract for Construction and General Conditions.

3. "Impact of Various Construction Types and Clauses on Project Performance," *Construction Institute Rep.* 10. (1986).

4. Schulman and Welby.

5. William S. McIntyre and Jack P. Gibson, *Construction Accounting Manual: Risk Management,* Chapter C4 (Warren Gorham & Lamont, 1996).

Chapter 9

1. Jerry Jackson, *Financial Management for Contractors,* 4th Ed. (FMI Corporation, 2002).

2. Steven D. Lord, *Cash Flow Formula* (Construction Financial Management Association, March–April 2003).

3. Dr. G. Howard Poteet, Comp., *Business Plan for Small Construction Firms/Starting Up Your Own Business: Expert Advice from the U.S. Small Business Association* (Liberty Hall Press,1991).

4. Ibid.

5. Jackson.

6. *How to Obtain Surety Bonds,* Surety Information Office, 5225 Wisconsin Avenue, NW, Suite 600, Washington DC 20015-2014, (202) 686-7463, *www.sio.org.*

THE STRATEGIC BUSINESS PLAN WORKBOOK

Company:

Address:

Telephone Number:

Contact:

Date:

EXECUTIVE SUMMARY

Suggestion: Write this section last. This is your chance to sell your company to the reader (owners, management team, potential lenders, and investors). First impressions are critical! Be concise and focused.

Purpose of Business Plan
Why are you writing one? As a management/operating tool? Do you want to get financing? What do you hope to accomplish?

Description of Business
What business are you in (i.e., general contractor, specialty contractor, supplier)? What is your service(s)? What markets do you serve (type of construction, type of client, geographic area)? What are the benefits to your user (quality, timeliness, etc.)? Stress particular areas of expertise and experience.

Target Markets
Who are your current customers/clients? Explain existing market and future potential.

Market Positioning
Describe the competitive environment and explain how your product/service is unique/better.

Strategic Direction

Explain your vision and how you plan to realize it. Outline: overall company goals and one-year goals, i.e., goals for growth, profitability, marketing, personnel development.

Management Team

Who are the important players? What is the background/experience/track record of each? Stress strengths of existing team and efforts being made to fill gaps.

Financial Factors

Summarize financial performance to date and explain your outlook based on financial projections. Establish goals for revenues, net income, return on investments, etc.

Financing Needs (if applicable)

How much financing is needed? What will it be used for? How will the funds be repaid and when? What can an investor expect in terms of annual return in the next three to seven years? Proposed investor payout? What other incentives are you offering?

Summary of Risks and Opportunities
Realistically assess the strengths and weaknesses inherent in your business. Explain the steps taken (to be taken) to mitigate risks.

BUSINESS DESCRIPTION

Company History
How long have you been in business? When and where company was founded, by whom, etc.

Progress Made to Date
Outline history of activities to date. What have been your accomplishments (and failures, if any) to date?

Company Legal Structure
Rationale for structure (sole proprietorship, partnership, corporation [C, S, LLC]). Joint ventures and other alliances, shareholder agreeements.

Status of MBE/WBE Certification (if applicable)
List state and city public agencies with which you are listed/active.

Overall Objectives/Mission Statement
State your mission clearly and simply. Where do you ultimately want to take the business? What is the philosophy that drives the company's overall goals?

Specific Objectives
Revenues, profitability, market share, product/work quality, technology, innovation, efficiency/productivity, management style, social responsibility, etc.

PRODUCTS AND SERVICES

Description
Be clear and specific.

User Benefits
What value/benefits do you provide to your clients? What problems/issues do you solve for them?

Proprietary Advantages (if applicable)
Patents, copyrights, trademarks, etc.

Government Permits, Licenses (if applicable)

MARKETS AND COMPETITION

Overall Trends in the Current Market
Location, size, economic factors, growth potential, leading players, etc.

Your Target Market
Location, size, growth trends, specific areas of opportunity. Who are the clients?

Competition
Who are they? How do they compare? Show that you know them. Who/what are your greatest threats?

Competitive Factors/Strategy
On what basis are you competing (expertise, pricing, quality, service, reliability, etc.)? How do you bid on and win new and repeat business? Do you have a market niche(s)?

General Market Influences
Economic, seasonal fluctuations, business and political environments, changing technology, random, etc.

Projected Revenues
One-year and two- to five-year scenarios. Show reasonable best/worst/most likely scenarios; explain assumptions clearly.

MARKETING

Results of Market Research
Research, surveys, feedback, presold units, etc., that determine short-term and long-term oppportunities in targeted areas.

Market Positioning/Image Portrayal
How do you come across? What is your motto? Careful positioning is important.

Target Audience
Owners, architects, government agencies, developers, brokers, etc.

Marketing Strategy

Formulate an appropriate mix for the near term and long term, including direct mail (letter campaign, brochures, business cards), bids/proposals, listings, advertising, public relations (articles, press releases), trade shows/exhibits, networking, etc. What is your annual budget?

Pricing Strategy

Review costs, desired profits, competition, image portrayal. What will the market bear?

Client Service/Follow-Up

Your approach to serving client needs and gaining repeat business.

Company Operations

Describe specific management approach and methodology. Outline company policies, procedures, systems, and controls. Stress advantages in production efficiencies, quality control, productivity.

Company Location

Benefits and drawbacks.

Construction Equipment/Facilities
Owned/leased? Rationale for what you have and what you need.

Organizational Structure
Create organizational chart of principals, managers, field personnel, and office personnel. Identify gaps that need to be filled.

Personnel Requirements and Strategy
Training, development, and performance evaluation for current employees. Hiring strategy for new hires. Outsourcing opportunities.

Project Bidding/Buyout/Planning Strategy
Describe how the project selection, bidding, and buyout processes are organized and managed. Discuss approach to site logistics, scheduling, cost engineering, project staffing, etc.

Risk Management
General liability, professional liability, life insurance, bonding, etc. Company's safety program and record. Pending claims/litigation. Disaster recovery plans.

Company Policies and Procedures

Describe what is in place and what needs to be established and implemented (office and field operations, record keeping, documentation, etc.).

Company and Project Controls/Systems

Describe existing project productivity and cost controls, company internal accounting controls and financial reporting, and quality and safety procedures and controls that exist and/or must be established. How are computer systems and software used to facilitate efforts? What are the gaps?

Contract Management

Explain procedures for monitoring contracts and managing/resolving claims/disputes.

MANAGEMENT TEAM

Key Players

Principals, key field personnel, key office personnel.

Background, Experience, Roles

Show how team members complement each other, work well together. Indicate those responsible for business development, general business management, marketing, and financial management.

Directors, Outside Advisors

Outside expertise who compensate for weaknesses in the management team (i.e., CPAs, attorneys, consultants, mentors).

FINANCIAL HISTORY AND REQUIREMENTS

Highlights of Historical Financial Statements

Explain trends in revenues, expenses, profits, balance sheet ratios. Compare with industry norms (i.e., Dun & Bradstreet).

Financial Goals

Outline your revenue, profit, and balance sheet goals for the next year and for the next five years.

Financing Needs (if applicable)

How much financing is needed? What will it be used for? How will the funds be repaid and when? What can an investor expect in terms of annual returns?

Highlights of Financial Projections

One-year and two- to five-year operating budgets and cash-flow projections. Explain perceived trends and overall direction. Have a list of assumptions in the appendix, along with spreadsheets.

Appendixes

Financial Statements	Customer References
Financial Projections	Contracts, Leases
Personal Financial Statements	Insurance Policies
Personal Guarantees	Bonding Capacity
Project Experience	Résumés of Key Personnel

B

INDUSTRY RESOURCES AND INFORMATION

Industry News and Information

Business.com	*www.business.com/directory/*
Business Publications Search	*http://bpubs.tradepub.com*
Construction Channel	*www.constructionchannel.com*
Construction Executive	*www.constructionexecutive.com*
Construction ProNews	*www.constructionpronews.com*
Construction Weblinks	*www.constructionweblinks.com*
Dodge Analytics	*www.mag.fwdodge.com*
Environmental Building News	*www.ebuild.com*
Info.com/construction	*www.info.com/construction*
InterPRO Resources	*www.ipr.com*
Reed Construction Data	*www.reedconstructiondata.com*
RS Means	*www.rsmeans.com*
Stamats Business Media	*www.buildings.com*
Thomas Register Online	*www.thomasnet.com*

Business Listings

Construction Data Company   *www.cdcnews.com*
McGraw-Hill Construction Network   *www.construction.com*
The Blue Book   *www.thebluebook.com*

Bidding Opportunities

Contractor's Register The Blue Book   *www.thebluebook.com*
BidClerk   *www.bidclerk.com*
McGraw-Hill's F.W. Dodge Reports   *www.fwdodge.com*
Reed Bulletin Construction Data   *www.cmdbulletin.com*

Periodicals

Building Design and Construction   *www.bdcmag.com.*
Construction Business Owner   *www.constructionbusinessowner*
                                         .com

Contracting Business   *www.contractingbusiness.com*
Contractor   *www.contractormag.com*
Engineering News-Record   *www.enr.com*
Free Trade Magazine Source   *http://af.freetrademagazinesource*
                                         .com

Jobsite Magazine   *www.jobsitemagazine.com*
McGraw-Hill Construction News   *www.construction.com*
Professional Builder   *www.probuilder.com*
Traditional Building   *www.traditional-building.com*

Construction Books

Bookmark Inc.   *www.bookmarki.com*
Construction Book Express   *www.constructionbook.com*
Contractor City Builders Bookshop   *www.contractorcity.com*

Training and Education

ASC Journal of Construction Education	*www.ascjournals.com*
Construction Industry Institute	*www.construction-institute.org*
Lorman Construction Seminars	*www.lorman.com*
OSHA Safety Training	*www.osha.gov*
Project Management Institute	*www.4pm.com*
RedVector	*www.redvector.com*

Other

A.M. Best Insurance Information Source	*www.ambest.com*
American Arbitration Association	*www.adr.org*
American Institute of Architects	*www.aia.org*
Construction Specifications Institute	*www.csinet.org*
National Safety Council	*www.nsc.org*

C

GOVERNMENT RESOURCES

Federal Government

FedBizOpps	*www.cbd-net.com*
Minority Business Development Agency	*www.mbda.gov*
U.S. Business Advisor	*www.business.gov*
U.S. Census Bureau	*www.census.gov*
U.S. Chamber of Commerce	*www.uschamber.org*
U.S. Department of Labor	*www.bls.gov/iag/construction .htm*
U.S. Internal Revenue Service	*www.irs.ustreas.gov*
U.S. Patent and Trademark Office	*www.uspto.gov*

U.S. Small Business Administration

www.sba.gov

Government Contracting	*www.sba.gov/gcbd*
HUBZone Program	*www.sba.govhubzone*
Office of Women's Business Ownership	*www.onlinewbc.gov*
Online Courses	*www.sba.gov/training*
Service Corps of Retired Executives	*www.score.org*

Small Business Development Centers	*www.sba.gov/sbdc*
Small Business Investment Companies	*www.sba.gov/INV*
Small Business Start-up Kit	*www.sba.gov/starting_business/ index.html*
Small Disadvantaged Businesses	*www.sba.gov/sdb*
8(a) Certification	*www.sba.gov/8abd*

D

MINORITY AND WOMEN BUSINESS ORGANIZATIONS

For Minority-owned Businesses

National Association of Minority Contractors *www.namcline.org*
National Minority Business Council *www.nmbc.org*
National Minority Supplier Development *www.nmsdcus.org*
   Council

For Women-owned Businesses

National Association of Women in *www.nawic.org*
   Construction
Women's Business Enterprise National *www.wbenc.org*
   Council
Women Construction Owners & Executives *www.wcoeusa.org*
Women Contractors Association *www.womencontractors.org*

INDUSTRY ASSOCIATIONS

Air Conditioning Contractors of America	*www.acca.org*
Alliance of Construction Trades	*www.actaz.net*
American Boiler Manufacturers Association	*www.abma.com*
American Ceramic Society	*www.acers.org*
American Concrete Institute	*www.aci-anti.org*
American Concrete Pavement Association	*www.pavement.com*
American Concrete Pipe Association	*www.concrete-pipe.org*
American Concrete Pressure Pipe Association	*www.acppa.org*
American Institute of Constructors	*www.aicnet.org*
American Institute of Steel Construction	*www.aisc.org*
American Institute of Timber Construction	*www.aitc-glulam.org*
American Iron & Steel Institute	*www.steel.org*
American Society of Concrete Contractors	*www.ascconc.org*
American Subcontractors Association	*www.asaonline.com*
American Underground Construction Association	*www.auaonline.org*
American Welding Society	*www.aws.org*
American Wood Council	*www.awc.org*
Architectural Woodwork Institute	*www.awinet.org*
Asphalt Institute	*www.asphaltinstitute.org*

Associated Builders & Contractors	*www.abc.org*
Associated General Contractors of America	*www.agc.org*
Associate Specialty Contractors	*www.assoc-spec-con.org*
Canadian Construction Association	*www.cca-acc.com*
Confederation of International Contractors' Associations	*www.cica.net*
Construction Estimating Institute	*www.estimating.org*
Construction Financial Management Association	*www.cfma.org*
Construction Industry Institute	*www.construction-institute .org*
Construction Industry Manufacturers Association	*www.cimate.com*
Construction Industry Round Table	*www.cirt.org*
Construction Innovation Forum	*www.cif.org*
Construction Management Association of America	*www.cmaanet.org*
Construction Owners Association of America	*www.coaa.org*
Construction Safety Council	*www.buildsafe.org*
Construction Specification Institute	*www.csinet.org*
Construction Suppliers Association	*www.gocsa.com*
Drywall Finishing Council Incorporated	*www.dwcf.org*
Edison Electric Institute	*www.eei.org*
HVAC Excellence	*www.hvacexellence.org*
Independent Electrical Contractors	*www.ieci.org*
Institute for Research in Construction	*www.cisti.nrc.ca/irc*
Institute of the Ironwork Industry	*www.instituteiw.org*
Insulation Contractors Association of America	*www.insulate.org*
International Assoc. of Foundation Drilling	*www.adsc-iafd.com*
International Assoc. of Plumbing & Mechanical Officials	*www.iapmo.org*
International Bridge, Tunnel & Turnpike Association	*www.ibtta.org*
International Builders Exchange Executives	*www.ibee.org*
International Construction Information Society	*www.icis.org*

International Facility Management Association	*www.ifma.org*
International Insulation Contractors Organization	*www.associationhouse.org*
International Masonry Institute	*www.imiweb.org*
International Project Management Association	*www.ipma.ch*
International Risk Management Institute	*www.irmi.com*
Land Improvement Contractors of America	*www.lica.org*
Mason Contractors Association of America	*www.masoncontractors.com*
Masonry Advisory Council	*www.maconline.org*
Masonry Institute of America	*www.masonryinstitute.org*
Mechanical Contractors Association of America	*www.mcaa.org*
Metal Building Contractors & Erectors Association	*www.mbcea.org*
NACORE International	*www.nacore.com*
National Alliance for Fair Contracting	*www.faircontracting.org*
National Asphalt Pavement Association	*www.hotmix.org*
National Association of Demolition Contractors	*www.demolitionassociation .com*
National Association of Elevator Contractors	*www.naec.org*
National Association of Home Builders	*www.nahb.org*
National Association of Plumbing, Heating & Cooling Contractors	*www.naphcc.org*
National Association of Reinforcing Steel Contractors	*www.narsc.com*
National Association of Remodeling Industry	*www.nari.org*
National Association of Women in Construction	*www.nawic.org*
National Association of Asphalt Technology	*www.ncat.us*
National Association of Masonry Association	*www.ncma.org*
National Assoc. of States on Building Codes & Standards	*www.ncsbcs.org*
National Contract Management Association	*www.ncmahq.org*
National Corrugated Steel Pipe Assoc.	*www.ncspa.org*

National Electrical Contractors Association	*www.necanet.org*
National Fire Protection Association	*www.nfpa.org*
National Glass Association	*www.glass.org*
National Guild of Professional Paperhangers	*www.ngpp.org*
National Institute of Building Sciences	*www.nibs.org*
National Insulation Association	*www.insulation.org*
National Lumber & Building Materials Dealers	*www.dealer.org*
National Paint & Coatings Association	*www.paint.org*
National Precast Concrete Association	*www.precast.org*
National Railroad Construction & Maintenance Association	*www.nrcma.org*
National Ready Mixed Concrete Association	*www.nrmca.org*
National Roofing Contractors Association	*www.nrca.net*
National Safety Council	*www.nsc.org*
National Stone Sand & Gravel Association	*www.nssga.org*
National Utility Contractors Association	*www.nuca.com*
National Wood Flooring Congress	*www.woodfloors.org*
New York Building Congress	*www.buildingcongress.com*
Pile Driving Contractors Association	*www.piledrivers.org*
Plumbing-Heating-Cooling Contractors Association	*www.phccweb.org*
Professional Construction Estimators Association	*www.pcea.org*
Professional Women in Construction	*www.pwcusa.org*
Professional Management Forum	*www.pmforum.org*
Professional Management Institute	*www.pmi.org*
Retail Contractors Association	*www.retailcontractors.org*
Risk & Insurance Management Society	*www.rims.org*
Scaffold Industry Association	*www.scaffold.org*
Sealant, Waterproofing & Restoration Institute	*www.swrionline.org*
Sheet Metal & Air Conditioning Contractors	*www.smacna.org*
Sheet Metal Workers' International Associations	*www.smwia.org*
Society of Cost Estimating & Analysis	*www.sceaonline.net*

Specialty Steel Industry of North America	*www.ssina.com*
Steel Construction Institute	*www.steel-sci.org*
Steel Erectors Association of America	*www.seaa.net*
Surety Association of America	*www.surety.org*
Surety Information Office	*www.sio.org*
The Masonry Society	*www.masonrysociety.org*
Tile Contractors Association of America	*www.tcaainc.org*
Timber Frame Business Council	*www.timberframe.org*
U.S. Green Building Council	*www.usgbc.org*
Urban Land Institute	*www.uli.org*
Washington Building Congress	*www.wbcet.org*
Western Building Material Association	*www.wbma.org*
Western States Roofing Contractors Association	*www.wsrca.com*
Western Wall & Ceiling Contractors Association	*www.wwcca.org*
Western Wood Products Association	*www.wwpa.org*

F

CONSTRUCTION UNIONS

AFL-CIO Building & Construction Trades Department	*www.buildingtrades.org*
International Association of Bridge, Structural, Ornamental and Reinforcing Iron Workers	*www.ironworkers.org*
International Association of Heat and Frost Insulators and Asbestos Workers	*www.insulators.org*
International Brotherhood of Electrical Workers	*www.ibew.org*
International Brotherhood of Teamsters	*www.teamster.org*
International Union of Bricklayers and Allied Craftworkers	*www.bacweb.org*
International Union of Elevator Constructors	*www.iuec.org*
International Union of Operating Engineers	*www.iuoe.org*
International Union of Painters and Allied Trades	*www.ibpat.org*

Laborers' International Union of North America	*www.liuna.org*
Operative Plasterers' and Cement Masons' International Association of the United States and Canada	*www.opcmia.org*
Sheet Metal Workers International Association	*www.smwia.org*
United Association of Journeyman and Apprentices of the Plumbing and Pipe Fitting Industry of the United States and Canada	*www.ua.org*
United Brotherhood of Carpenters and Joiners of America	*www.carpenters.org*
United Union of Roofers, Waterproofers and Allied Workers	*www.unionroofers.com*
Utility Workers Union of America	*www.uwua.org*

SAMPLE PROJECT MANAGEMENT FORMS AND TEMPLATES

Sample Cost Estimate

COMPANY
Project XYZ

CSI Code	ITEM	DESCRIPTION	BREAKDOWN				Adjustments	Comments
			Quant	Unit	Unit Price	Total		
		SUBTOTAL, TRADE COST ==============>>>>>						
		GENERAL CONDITIONS @ 5 -10% (depending on job conditions and size)						
		INSURANCE & BOND						
		NEW SUBTOTAL ============>>>>>						
		Overhead and Profit @ %						
		DESIGN CONTINGENCY @ %						
		GRAND TOTAL ============>>>>>						

COMPANY
Contractor Qualification Form

It is our policy, before we use quotes or sign contracts, to ask contractors to submit this qualification form. This enables us to categorize contractors within their trade by types and sizes of contracts they can handle.

Please complete the form and submit the following attachments with it:
- Financial statements (copies of your three most recent annual financial statements);
- Licenses (copies of your current license or certification, if you are an electrician, plumber, asbestos handler, or in any other trade that requires a license or certification to perform work);
- Resumes (copies of the resumes of all of your key people—that is, officers, partners, owners, and managers with experience in the type of work for which you seek qualification).

1. Contractor Identity

Area of expertise_____

Company name _____

Address_____

Phone # _____ Fax # _____ E-mail address _____

Tax ID or SS # _____ Contact person _____

Type of company: ❑ Sole proprietorship ❑ Corporation ❑ Partnership ❑ Date formed _____

States in which the company is legally qualified to do business _____

Total number of employees_____

Names and titles of key people in company _____

_____

Has the company operated under any other name in the past five years? ❑ yes ❑ no

If yes, give name(s) _____

Does the company have offices, plants, or warehouses at other locations? ❑ yes ❑ no

If yes, list addresses_____

2. MBE/WBE/SBE Certification

Is the company a certified Minority Business Enterprise (MBE), Women Business Enterprise (WBE), Small Business Enterprise (SBE), or any other type of certified business enterprise? ❑ yes ❑ no

If yes, list. _____

_____

3. Bank Reference

Does the company have a line of credit from any lending institution? ❑ yes ❑ no If yes, give details:

AMT. OF CREDIT   OUTSTANDING BALANCE   LENDER'S NAME/ADDRESS   LENDING OFFICER'S NAME/PHONE #

_____

_____

_____

4. Bonding Capacity

Do you have bonding? ❑ yes ❑ no If yes, give details:

Single project limit _____ Aggregate limit _____

Bonding company name/address _____

Bonding agent name/address /phone #_____

5. Completed Projects (Summarize representative projects completed in past five years)

NAME OF PROJECT	SCOPE OF WORK	CONTRACT AMOUNT	COMPLETION DATE

6. Current Projects (Summarize current projects)

NAME OF PROJECT	SCOPE OF WORK	CONTRACT AMOUNT	COMPLETION DATE

7. Trade References (List three of your subcontractors or suppliers)

NAME OF PROJECT	SCOPE OF WORK	CONTRACT AMOUNT	COMPLETION DATE

8. Client References (List three clients)

NAME OF PROJECT	SCOPE OF WORK	CONTRACT AMOUNT	COMPLETION DATE

9. Other Information

Has your company or any of its key people been a party to a bankruptcy or reorganization proceeding?
❑ yes ❑ no
If yes, give date. _____

During the past five years, have any liens been filed against you by any of your subcontractors or suppliers?
❑ yes ❑ no If yes, give details for any liens over $5,000. _____
_____

Have you ever failed to complete a contract, been defaulted, or had a contract terminated? ❑ yes ❑ no
If yes, give details. _____
_____
_____

In the past five years, have you had liquidated damages assessed against you upon completion or a project?
❑ yes ❑ no If yes, give details. _____
_____

In the past five years, has your company or any of its key people been involved in any lawsuits arising from
construction projects? ❑ yes ❑ no If yes, give details. _____
_____

In the past five years, has your company or any of its key people been investigated for or found to have committed
a violation of any labor laws? ❑ yes ❑ no If yes, give details. _____
_____

In the past five years, has your company or any of its key people been investigated for or found to have committed
a serious OSHA violation? ❑ yes ❑ no If yes, give details. _____
_____

In the past five years, has your company or any of its key people been investigated for or found to have committed
a violation of state, federal, or local environmental protection laws? ❑ yes ❑ no
If yes, give details. _____
_____

Is there any other information you would like to give us? _____
_____
_____
_____

COMPANY
Telephone Quotation

Subcontractor _____ By _____

Trade _____ Tel. No. _____

Job _____

Amount, Plans & Specs. _____ Erected or   Time to

F.O.B. _____ Complete_____

Sales Tax Incl. ❑ yes ❑ no   Bondable ❑ yes ❑ no   Insurance Incl. ❑ yes ❑ no

Addenda Incl. _____ Nos. _____ Hoisting Incl. ❑ yes ❑ no

(Check Above By Reading Back)

_____
_____
_____
_____
_____
_____
_____
_____
_____
_____
_____
_____
_____
_____
_____

Exceptions:_____
_____
_____
_____
_____
_____
_____

Alternates: _____
_____
_____
_____

This quotation was received by _____ Estimating Department

Date quotation was received _____ 20____ ____ AM ____ PM

COMPANY
Subcontractor/Supplier Quote Information

Project Information
Project name _____
Bid solicitation # _____ Date/Time of bid opening _____

Subcontractor/Supplier Information
Name_____
Representative's name/title_____
Phone # _____ Fax # _____
Is subcontractor licensed? ❏ yes ❏ no
Type of license? _____ License # _____
Is subcontractor a certified minority contractor? ❏ yes ❏ no Circle type: **MBE WBE DBE SBE**
Certifying agency _____ Certification #_____
Is subcontractor a union contractor? ❏ yes ❏ no Identify union _____
Is subcontractor bondable? ❏ yes ❏ no What is the bond premium rate? _____

Quote Information
This quote is based on plans and specs dated _____
Is sub/supplier quoting as per plans and specs? ❏ yes ❏ no
Quote reflects the following addenda numbers _____
Scope of Work *(Give specific sections, work description, and corresponding quote amounts below.)*

Section	**Description**	**Amount (+ or -)**
_____	_____	_____
_____	_____	_____
_____	_____	_____

Alternate #	**Description**	**Amount (+ or -)**
_____	_____	_____
_____	_____	_____
_____	_____	_____

Are there any exclusions or clarifications? ❏ yes ❏ no
If yes, describe. _____
_____
Does quote include delivery to job sites? ❏ yes ❏ no
Does quote include installation? ❏ yes ❏ no
Does quote include sales tax? ❏ yes ❏ no
If permit fees are associated with this work, are they included in quote? ❏ yes ❏ no
Do you have any ideas for savings related to this work? _____
_____
Information received by _____

COMPANY
REQUEST FOR INFORMATION

TO:	R.F.I #:
Company:	Date:
RE:	Respond By:
CC:	Drawing No.:

Question:

Please respond in writing TO:	Addressed By: Signature: Date:

Answer:

Activity ID	Activity Description	Orig Dur	Rem Dur	Early Start	Early Finish	Actual Start
TEMPORARY CERTIFICATE OF OCCUPANCY						
2080	New Elevator Furnishing & Installation	70	8	07OCT00A	23FEB01	07OCT00
2030	Concrete & Masonry Work	40	4	25OCT00A	26FEB01	25OCT00
2000	Obtaining General Building Permit	0	0	25OCT00A		25OCT00
2010	Obtaining Fire Alarm Building Permit	0	0	06DEC00A		06DEC00
2280	New MEP Engineering Design & Sketches	5	0	06DEC00A	15JAN01A	06DEC00
2090	Structural Steel Work @ Roof & Stair #1	15	0	15DEC00A	01JAN01A	15DEC00
2290	Remedial HVAC Work	15	8	05JAN01A	05MAR01	05JAN01
2120	Patch Roof	5	0	12JAN01A	18JAN01A	12JAN01
2320	Delays to Concrete Wrk due to Field &	10	5	02FEB01A	20FEB01	02FEB01
2370	Connect & Power Scoreboard	1	1	14FEB01	14FEB01	
2300	Remove and Reinstall Wall Padding	40	28	14FEB01*	23MAR01	
2160	Power and Install Cntrls for Basket. Motors	10	10	14FEB01	27FEB01	
2070	Remedial Electrical Work & Inspector's	5	5	14FEB01	27FEB01	
2100	Furnish & Install New Handrails	15	15	22FEB01	14MAR01	
2350	Cleaning of Gym Interiors and Equipment	15	15	01MAR01	21MAR01	
2170	Remedial Fire Alarm Work	2	2	07MAR01	08MAR01	
2180	Furnish & Install New Terrazzo	30	30	15MAR01	25APR01	
2190	Drywall and misc. Patch & Paint	15	15	05APR01	25APR01	
2150	Sign-off on Structural, MEP & Architectural	5	5	26APR01	02MAY01	
2250	Certificate Of Occupancy	0	0		02MAY01	
FIRE PUMP						
2270	New MEP Engineering Design & Sketches	5	0	20NOV00A	01DEC00A	20NOV00
2220	Complete Power Hook-up	2	0	05DEC00A	06DEC00A	05DEC00
2260	Complete Remaining Mechanical Work	2	0	15JAN01A	17JAN01A	15JAN01
2210	Fire Alarm Work	2	0	17JAN01A	17JAN01A	17JAN01
2330	Repair Fire Main Leak	4	1	12FEB01A	14FEB01	12FEB01
2360	Complete Inspector's Checklist	5	5	14FEB01	20FEB01	
2380	Delays due to Fire Main Leak	10	10	14FEB01	27FEB01	
2230	Test & Start-up	1	1	22FEB01*	22FEB01	
2240	Final Inspection	0	0	22FEB01*	22FEB01*	
OWNER UPGRADE						
2110	Revise the Balcony Extension	35	20	02FEB01A	13MAR01	02FEB01
2130	Construct New Driveway	30	30	03MAY01	13JUN01	
2340	Furnish & Install New Trench Drain	10	10	03MAY01	16MAY01	
2140	Owner Upgrade Complete	0	0		13JUN01	
PROJECT CLOSE-OUT						
2310	Project Complete	0	0		02MAY01	

Chart annotations:

Delays as per SoundBuild let 12-22-00 and 01-19-01.

Start date for terrazzo work may be re-established. Pending design on Stair #4.

Fire pump test schedule may be re-established. Depending Completion of Inspector's Checklist.

Project Start	28SEP00	
Project Finish	13JUN01	Early Bar
Data Date	14FEB01	Progress Bar
Run Date	07SEP01	Critical Activity

GSO4

SoundBuild, Inc. Management Plan
German School of NY Corrective Work
Certificate of Occupancy & Beyond

Sheet 1 of 1

© Primavera Systems, Inc.

COMPANY

Project XYZ

Two Week Look Ahead Schedule

Job Name: _____

Period: _____

DATE: _____

Activity	MO	TU	WE	TH	FR	SA	SU	MO	TU	WE	TH	FR	SA	SU	Comments

COMPANY

TRANSMITTAL

TO:	DATE:
FROM:	RE:

WE ARE SENDING YOU THE FOLLOWING VIA:

COPIES	DWG NO.	DATE	DESCRIPTION

THE ABOVE IS BEING SENT TO YOU FOR THE FOLLOWING REASON:

REMARKS:

Shop Drawings Log
Project XYZ

COMPANY

Updated on:
Print Date:

Spec. Section	Subm. No.	Drwg. No.	Drawing Description / Title	Date of Drwg.	Date Received	Date to A/E	Received fr. A/E	Days @A/E	Action	Return to Sub.	Resubmit Y / N	Comments

Payment Log
Project XYZ

COMPANY

Status as of:

Contract Sum:

Requisition #	Bill		Payment	
	Date	Amount	Date	Amount

Total   0   Total   0
Remaining Total   0

Change O.#	Bill		Payment	
	Date	Amount	Date	Amount

Total   0   Total   0
Remaining Total   0

RFI Log
Project XYZ

COMPANY

Status as of:

RFI#	Subject	Trade	Date Issued	Priority	Date Resolved	Days Outstanding	Comments

COMPANY

Contract Drawings Log

Project XYZ

Updated as of (Date)

Dwg No.	Revision	Description	Dated	Date Received

	COMPANY
	PROGRESS REPORT

PROJECT XYZ

Report Date:
Print Date:

Time Period: 123
To:
Report Prepared By:

Trade	Description	% Complete	Activity

Problems/Corrective Actions/Projections:

_____
Project Manager

_____
Project Superintendent

AIA Document G702, APPLICATION AND CERTIFICATE FOR PAYMENT,
containing Contractor's signed Certification is attached.
In tabulations below, amounts are stated to the nearest dollar.
Use Column I on Contracts where variable retainage for line items
may apply.

APPLICATION NUMBER:
APPLICATION DATE:
PERIOD TO:
CONTRACTOR'S PROJECT NO:
CONTRACTOR'S NAME:

A	B	C	D	E	F	G	%	H	I
			WORK COMPLETED						
ITEM NO.	DESCRIPTION OF WORK	SCHEDULED VALUE	PREVIOUS APPLICATION (D+E)	WORK IN PLACE THIS PERIOD	STORED MATERIALS (NOT IN D OR E)	TOTAL COMPLETED AND STORED TO DATE	CPLT (G/C)	BALANCE TO FINISH (C-G)	RETAINAGE 0%
TOTALS									

COMPANY

SUMMARY PROJECT COST REPORT
Project XYZ
Address
Month

Cost Code	Description	Original Budget	Budget Change Orders			Revised Budget	Payments To Date	To Be Paid Including Retainage	% Work To Date
			Approved	Submitted	Pending				
Subtotal									
GRAND TOTAL									
DESIGN CONTINGENCY									
CONSTRUCTION CONTINGENCY									
PROJECT TOTAL									

BUDGET COSTS

COMPANY

<u>EMERGENCY NUMBERS</u>

ALL ACCIDENT & FIRE EMERGENCIES:

LOCAL POLICE: _____

      Contact: _____

FIRE DEPARTMENT: _____

      Contact: _____

AMBULANCE: _____

      Contact: _____

LOCAL HOSPITAL: _____

      Contact: _____

PROJECT OFFICIAL: _____

INFORMATION TO SUBMIT TO WHEN REPORTING AN EMERGENCY:

      Nature of Emergency

      Site Location

      Name of Person Calling

      Name of person(s) injured and/or apparent damages

Information for Off-Hours Personnel:

      Location of Phone: _____

COMPANY

Emergency Contact List

NAME	TITLE/COMPANY	OFFICE NO.	BPR/CELL	HOME NO.	COMMENTS

SAMPLE TABLE OF CONTENTS
FOR EMPLOYEE MANUAL

SECTION 1: THE WAY WE WORK

> TYPES OF SERVICES WE OFFER
> EQUAL EMPLOYMENT OPPORTUNITY
> AMERICANS WITH DISABILITIES ACT
> AIDS IN THE WORKPLACE
> A WORD ABOUT OUR EMPLOYEE RELATIONS PHILOSOPHY
> NONHARASSMENT
> SEXUAL HARASSMENT
> CATEGORIES OF EMPLOYMENT
> NEW EMPLOYEE ORIENTATION
> SUGGESTIONS AND IDEAS
> TALK TO US

SECTION 2: YOUR PAY AND PROGRESS

> RECORDING YOUR TIME
> PAYDAY
> METHOD OF COMPENSATION
> PERFORMANCE REVIEWS

PROMOTIONS
PAY RAISES
OVERTIME

SECTION 3: TIME AWAY FROM WORK AND OTHER BENEFITS

HOLIDAYS
VACATION
SICK DAYS
JURY DUTY
VOTING LEAVE
MILITARY LEAVE
WITNESS LEAVE
BEREAVEMENT
LEAVE OF ABSENCE
MEDICAL INSURANCE
DENTAL INSURANCE
COBRA
SHORT-TERM DISABILITY
FEDERAL FAMILY AND MEDICAL LEAVE
LONG-TERM DISABILITY
SOCIAL SECURITY
WORKERS' COMPENSATION
INCENTIVE PLANS
SECTION 125 PLAN
PROFESSIONAL DEVELOPMENT
TUITION REIMBURSEMENT

SECTION 4: ON THE JOB

VOICE MAIL
E-MAIL
PERSONAL TELEPHONE CALLS
ATTENDANCE AND PUNCTUALITY
WORKWEEK
MEALTIME
WORK ASSIGNMENTS

SECTION 5: SAFETY IN THE WORKPLACE

SAMPLE TABLE OF CONTENTS FOR PROJECT MANAGEMENT MANUAL

1. **General Policies**
   a. Company Mission Statement
   b. Company Policy Toward Projects
   c. Company Policy Toward Clients

2. **Project Management Team**
   a. Job Description for Project Manager
   b. Job Description for Superintendent
   c. Job Description for Foreman

3. **Project Bidding/Buyout**
   a. Bid/No Bid Decision
   b. Site Investigation Checklist
   c. Cost Estimate
   d. Bid Form
   e. Materials List
   f. Subcontractor Prequalification
   g. Vendor Quote Information
   h. Subcontractor Bid Review

     i. Bid Tabulation Sheet

     j. Construction Start Items

     k. Long Lead Items

     l. Bid Proposal

4. **Project Planning/Scheduling**
   a. Permits
   b. Manpower Planning
   c. Site Logistics Plan
   d. Mobilization Plan
   e. Baseline Project Schedule
   f. Two-Week Look-Ahead Schedule
   g. Project Kick-Off Meeting
   h. Project Directory

5. **Contracts**
   a. Construction Contracts
   b. Important Contract Provisions
   c. Sample Subcontractor Contract
   d. Purchase Orders
   e. Insurance Requirements

6. **Submittals**
   a. Submittal Schedule
   b. Submittal Log
   c. Submittal Form
   d. Request for Information Form
   e. Request for Information Log

7. **Changes**
   a. Change Order Procedures
   b. Change Order Request Form
   c. Field Orders
   d. Time and Materials Tickets
   e. Change Order Status Log

8. **Manpower/Payroll**
   a. Daily Construction Report
   b. Weekly Time Sheets
   c. Payroll Reports

9. **Payment Requests**
   a. Contract Payment Provisions
   b. Schedule of Values
   c. Payment Procedures
   d. Payment Requisition Form
   e. Subcontractor Approval for Payment
   f. Waiver and Release of Lien

10. **Cost Reporting**
    a. Cost-Reporting Procedures
    b. Project Cost Report
    c. Cash-Flow Projections

11. **Progress Meetings/Reports**
    a. Project Progress Meeting Procedures
    b. Meeting Minutes
    c. Progress Reports

12. **Correspondence/Filing**
    a. Filing System
    b. Standardized Letters

13. **Quality Control**
    a. Quality-Control Checklist
    b. QA Log

14. **Safety**
    a. Safety Inspection Procedures
    b. OSHA Inspection Instructions
    c. Hazard Communication Standard
    d. Safety Program
    e. Safety Training
    f. Safety Meetings
    g. Job Site Safety Talks
    h. Safety Equipment
    i. Emergency Contact List
    j. Accident Report Form

15. **Closeout**
    a. Punch-list Procedures
    b. Correspondence
    c. Turnover of Systems

SAMPLE PROJECT FILING SYSTEM

Number	Main Heading
00200	**Preconstruction**
00210	Preliminary Schedules
00220	Geotechnical Studies, Engineering Studies
00230	Project Site Plan
00240	Project Financial Information
00300	**Project Bidding**
00301	Bid Form/Alternates/Allowances/Qualifications and Exclusions
00302	Owner Contract and List of Contract Documents
00303	Project Estimate
00310	Project Addenda
00320	Subcontractor's Pricing
00500	**Contract**
00501	Letter of Intent/Notice to Proceed
00502	Executed Contract with Owner and Insurance and Bond Certificates
00505	Project Baseline Schedule
01025	**Payment Applications**
01025A	Schedule of Values
01025B	Payment Applications—by number

01035	**Change Orders/Modifications**
01035A	Change Order Requests
01035B	Owner Change Orders
01035C	Subcontractor Change Orders
01035D	Field Orders
01035E	Directives
01060	**Regulatory Requirements**
01061	Building Permits
01062	Inspections and Sign-offs (by trade)
01063	ACM Issues/Asbestos Containment Material
01065	TCO/CO
01200	**Meetings**
01210	Preconstruction Meeting
01220	Progress Meeting Minutes with Owner
01221	Progress Meeting Minutes with Subcontractors
01230A	Safety Meeting Minutes with Subcontractors
01230B	Toolbox Safety Meeting Minutes with MSDS Training
01245	Special Installation Meeting Minutes
01250	**Reports/Logs**
01251	Superintendent's Daily Reports
01252	Project Manager's Weekly Diary
01270	**Correspondence**
01270A	Correspondence to Owner
01270B	Correspondence from Owner
01270E	Correspondence to Architect
01270F	Correspondence from Architect
01270G	Correspondence to/from AE
01300	**Submittals**
01310	Progress Schedules—by month, week, etc.
01320	Progress and Cost Reports—by month
01330	Survey, Layout, and Initial Project Conditions
01340	Shop Drawings, Product and Data Samples
01380	Construction Progress Photos
01400	**Quality**
01405	Punch List
01410	Testing Laboratory Services
01420	Inspection Services
01425	Field Samples and Mockups

01430	Field Reports
01445	Manufacturer Field Services
01446	A/E Field Reports
01500	**Construction Facilities and Temporary Controls**
01505	Site Logistics Plan/Mobilization
01510	Temporary Utilities
01510-01	*Temporary Fire Protection*
01510-02	*Temporary Water and Sewer*
01510-03	*Temporary Heat*
01510-04	*Temporary Electric*
01510-05	*Temporary Utilities for Field Office*
01525	Construction Aids
01525-01	*Elevators*
01525-02	*Hoists*
01525-03	*Cranes*
01525-04	*Scaffolding*
01525-05	*Swing Stage*
01540	Security Service
01560	Temporary Controls
01560-01	*Site Cleaning*
01560-02	*Dust and Erosion Control*
01560-03	*Pest and Rodent Control*
01560-04	*Noise and Pollution Control*
01560-05	*Surface Water Control*
01580	Project Signs
01590	Field Offices and Sheds
01650	**Systems Startup/Commissioning**
01650A	Starting of Systems/Schedule
01650B	Systems Demonstration
01700	**Contract Closeout**
01710	Final Cleaning
01720	Project Record Documents
01730	Operation and Maintenance Data
01740	Warranties and Bonds
01750	Spare Parts/Attic Stock/Maintenance Materials
01900	**Other**
01910	Chronological File
01920	Legal

01930	Safety
01930-01	*Certifications*
01930-02	*Training*
01930-03	*MSDS*
01940	Violations
01950	Accidents
01960	**Subconsultants**
01961	**Sub #1**
01961A	Agreement and Insurance
01961B	Correspondence to/from
01961C	Payment and Invoices
01975	**Requests for Information**
01975	RFI Number Log (Number Order)
01975-01	RFI # 1
01980	**Sketches**
01980	Sketch Number Log (Number Order)
01990	**Bulletins**
01990	Bulletin Number Log (Number Order)
15000	**Subcontractors—Mechanical**
15000-01	Buyout
15000-02	Subcontract, Insurance and CO's File
15000-03	Correspondence from Subcontractor
15000-04	Correspondence to Subcontractor
15000-05	Notices and Back-Charges

PROJECT MANAGEMENT PROCEDURES WORKBOOK

PROJECT BIDDING/BUYOUT

Bidding efforts should be carefully targeted and organized to minimize risks down the road.

❑ **What is the PM's involvement in the bidding/buyout process?**

Comments:

Recommendations:

❑ **What procedures are used to review bid documents, understand the scope of work, uncover particular site conditions and design/ scope discrepancies, and prepare cost estimates?**

Comments:

Recommendations:

❏ **How are subcontractors/consultants selected/prequalified?**

Comments:

Recommendations:

❏ **What are the mechanics of collecting/disseminating plans to subs and follow-ups?**

Comments:

Recommendations:

❏ **How is cost/value engineering performed to identify potential cost/time savings?**

Comments:

Recommendations:

❏ **How is the bid tabulation process executed to prepare a bid and incorporate VE, site conditions, and discrepancies in documents, etc.?**

Comments:

Recommendations:

PROJECT PLANNING

Proper planning is critical to project success.

❑ **How is the project team selected?**

Comments:

Recommendations:

❑ **Are there clearly defined roles and responsibilities and key performance criteria for the project team members?**

Comments:

Recommendations:

❑ **How are communications channels developed (i.e., prejob conferences, kick-off meetings)?**

Comments:

Recommendations:

❑ **How is the project schedule prepared/updated and used as a management tool? How is the project weekly/biweekly look-ahead schedule prepared and used to manage subcontractor performance?**

Comments:

Recommendations:

❏ **How is the job mobilization plan implemented (i.e., site logistics)?**

Comments:

Recommendations:

CONTRACTS

Key project team members must know contracts with the customer and subs and manage accordingly.

❏ **How are contracts/subcontracts reviewed/written?**

Comments:

Recommendations:

❏ **How is the project team made fully aware of key provisions/potential pitfalls?**

Comments:

Recommendations:

❏ **How are strategies developed to mitigate problems?**

Comments:

Recommendations:

❑ **How are subcontracts negotiated and managed?**

Comments:

Recommendations:

❑ **How are change orders negotiated/managed?**

Comments:

Recommendations:

❑ **Is there a formalized policy for claims/disputes?**

Comments:

Recommendations:

POLICIES AND PROCEDURES/CONTROLS/ RECORD KEEPING

Company policies and procedures must be clear and consistent; proper documentation is crucial to requesting additional time/money.

❑ **Is there a current project management manual?**

Comments:

Recommendations:

❑ **How does information flow between the project site, the client, the design team, and the main office?**

Comments:

Recommendations:

❑ **Are there *standardized* forms for RFIs, transmittals, submittals, daily logs, time sheets, memoranda/letters, change orders, purchase orders, various logs, meeting minutes, etc.?**

Comments:

Recommendations:

❑ **How/where is day-to-day project documentation filed and tracked?**

Comments:

Recommendations:

❑ **How are adverse field conditions, delays, or other occurrences documented?**

Comments:

Recommendations:

❑ **Are there updated quality-control and safety policies and procedures?**

Comments:

Recommendations:

❑ **What procedures are used to ensure proper closeout and full retainage collection?**

Comments:

Recommendations:

❑ **What is the procedure for obtaining and maintaining proper insurance and bonding coverage?**

Comments:

Recommendations:

PROJECT CONTROLS AND REPORTING

Project team members must have a finger on the pulse of project costs and progress at all times.

❑ **How are project costs, revised budgets, and pending change orders tracked and compared with the original budget?**

Comments:

Recommendations:

❑ **Are project cash-flow forecasts and/or other PM forecasting tools used?**

Comments:

Recommendations:

❑ **How are payment applications managed?**

Comments:

Recommendations:

❑ **How is labor productivity gauged and managed?**

Comments:

Recommendations:

❑ **What are the procedures for schedule/progress updates (WIPs, hot items, various logs), job site meetings/minutes, and follow-up?**

Comments:

Recommendations:

❑ **How does information/feedback/direction flow back to the PM from senior management?**

Comments:

Recommendations:

SYSTEMS

Software packages are available to streamline paper flows and communication and standardize/integrate documentation.

❑ **What software is currently being used/by whom?**

Comments:

Recommendations:

❑ **How are staff members trained?**

Comments:

Recommendations:

❑ **Does the current system provide linkage between the project site and main office and between staff members?**

Comments:

Recommendations:

❑ **Does it link project cost and accounting data?**

Comments:

Recommendations:

❑ **Does it allow for easy access to standard forms and spreadsheets?**

Comments:

Recommendations:

❑ **How are completed projects archived for use as historical data or as backup for future claims/disputes?**

Comments:

Recommendations: